PHL
5406000006642

THE MOON

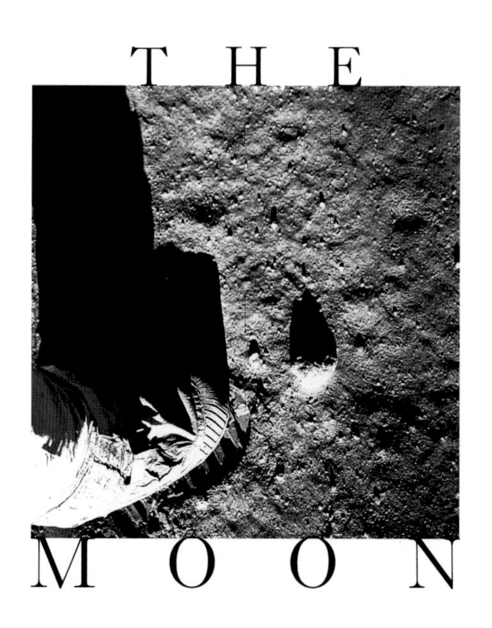

T H E

M O

O N

Maryam Sachs

ABBEVILLE PRESS PUBLISHERS
NEW YORK LONDON PARIS

Contents

Front cover: Sonja Bullaty, *Stonehenge,* 1969. Color photograph.
Page 1: *First Step on the Moon,* July 1969. NASA photograph.
Pages 2–3: Georges Méliès (1861–1938). *A Trip to the Moon,* 1902. Film still.
Page 4: Gunter Sachs (b. 1933). *Dawn,* 1980. Color photograph.
Pages 6–7: George Robbins (b. 1933), *Setting Moon in Early Morning, Melville, Montana,* n.d. Color photograph.

Editor: Abigail Asher
Translator: Alexandra Chapman
Designer: Celia Fuller
Production Editor: Owen Dugan
Picture Editors: Naomi Ben-Shahar, Scott Hall, and David Parket
Production Director: Hope Koturo

 Inquiries should be addressed to Abbeville Publishing Group, 22 Cortlandt Street, New York, N.Y. 10007. The text of this book was set in Caslon C37 and Moulin Rouge Solide. Printed and bound in China.

First edition
10 9 8 7 6 5 4 3 2 1

Library of Congress Cataloging-in-Publication Data
Sachs, Maryam, 1961–
The moon / by Maryam Sachs.
p. cm.
ISBN 0-7892-0341-3
1. Moon–Popular works. I. Title.
QB581.9.S2 1998
523.3–dc21 98-17962

To my sun, Jounam
To my stars, Philipp, Frederik, and Roya

FROM YOUR MOON

René Magritte (1898-1967).
Spring, 1965. Oil on canvas,
21 5/8 x 18 1/8 in. (55 x 46 cm).
PRIVATE COLLECTION.

Introduction

I must repeat that the distance between the Earth and its satellite is really not very great and hardly worthy of concern to serious minds. I therefore believe it is not too bold of me to predict that there will soon exist a series of projectiles providing comfortable transportation between the Earth and the Moon. There will be no risk of shocks, jolts, or derailing, and we will reach our destination rapidly, directly, and without fatigue. . . . Twenty years from now, half the population of the Earth will have visited the Moon!

JULES VERNE, *From the Earth to the Moon*

This optimistic statement was made by a character in the novel, one of the three passengers in the "bullet train" that undertook the first voyage to the Moon–only to miss its mark and end up as a lunar satellite. When his book was published in 1865, such a voyage was utter fantasy. Even picture-book hero Tintin, who dreamed of traveling to the Moon in *Objective Moon* and actually did so in *They Walked on the Moon,* was still some fifteen years ahead of history. The fact is that the idea of going to the Moon has always fascinated the human animal. As early as the second century, the Greek poet Lucianus of

Sigmar Polke (b. 1941).
Leave Me Alone, 1964. Drawing on paper, 11 1/8 x 8 1/8 in. (28.2 x 20.5 cm).
PRIVATE COLLECTION.

Samosata described the saga of a ship swept up to the Moon by a storm. In 1656, in *The Other World,* Cyrano de Bergerac relates how an airship, propelled by the evaporation of dew, transports him to a Moon that turns out to be quite similar to Earth. As for Hans Pfaall of Rotterdam, he reached the Moon in nineteen days by means of a balloon propelled by the pen of Edgar Allan Poe.

Now that this dream voyage has become reality, it is apparent that dreaming of going to the Moon and actually going there are two quite different things. One might even add that humanity is now divided into two groups: on the one hand, the immense majority of people whose relationship with the Moon is merely imaginary, and on the other hand, the few privileged individuals who have actually set foot on the Moon.

The emotional impact of the adventure has perhaps been overshadowed by the technological exploits involved, which are indeed fantastic. However, astronauts who have seen the Earth from the viewpoint of the Moon as a mere grain of dust in the universe must have undergone an emotionally shattering experience. On their return to Earth, these men must have had fascinating things to say; they may even have sketched amazing landscapes; and they certainly must have written

Mike Blake.
Harvest Moon, Toronto,
September 17, 1986. Photograph.

Norman Rockwell (1894-1978).
First Step on the Moon, 1966. Oil on canvas, 64 x 40 in.
(163 x 101.6 cm). NATIONAL AIR AND SPACE MUSEUM,
SMITHSONIAN INSTITUTION, WASHINGTON, D.C.

letters describing their impressions. But where is the record of their emotional reactions? Dazzled and impressed by technological data, the press limited its "human interest" reports to anecdotes about the freedom of weightlessness, for example. This is why I believe it is important to highlight the relationship between the Moon and humankind and their adventures throughout all ages and cultures. It is obviously too vast a subject for a single volume. I have simply focused on the aspects that seem to me particularly mysterious and intriguing.

The newspaper article that most impressed me as a child was headlined: MAN ON THE MOON. I was only eight years old at the time, but I can still see that paper on my parents' bed table. For children it was impossible to separate reality from fairytale in that fabulous event; to an eight-year-old, the boundary between imagination and reality is always rather vague, and the historical conquest of the Moon is easily confused in head and heart with all those charming tales of climbing to the Moon on a ladder. And to a child, after all, there is little difference between a parent telling a bedtime story and an astronaut telling an eyewitness story from the Moon.

At the turn of the millennium, humankind's grasp is greater, and Shakespeare's "silver bow new-bent in heaven" is slated for use as a fueling station and a base for rocket expeditions to Mars. What an imaginative exploitation of ice and water! As for me, I find that, despite its "conquest" by the Earth, the Moon is more mysterious than ever.

Fin du Monde
le 19 Mai 1910.
On s'en f...... p
Expédition
à la Lune.
Grande vitesse!
200 Frs. le coup.
Le dernier salut!

Moon Facts

Vital Statistics

NAMES: Anaitis, Annis, Anu, Aphrodite, Artemis, Bridgit, Cybele, Demeter, Diana, Hathor, Hecate, Ishtar, Khonsou, Luna, Min, Osiris, Phoebe, Selene, Shing-Moo, Sin, Thoth, etc.

BIRTHPLACE: Outer space

DESTINATION: Around the world in 27 days, 7 hours, 43 minutes, at an average speed of 2,300 miles (3,700 km) per hour

PRINCIPAL OCCUPATION: Reflecting the light of the Sun

NEXT OF KIN: Earth, which lives from 226,000 to 252,000 miles (363,000 to 406,000 km) away

DISTINCTIVE FEATURES: Absence of air, absence of wind, absence of life. Presence of water recently proven

PLANTS: The American flag

COLOR: Silver, red, orange, brown, blue

DIAMETER: 2,160 miles (3,476 km)

WEIGHT: Six times less than on Earth

HEIGHT: 5 miles (8.2 km) at its highest peak

TEMPERATURE: Depending on the sunshine, from 212°F to -200°F (100°C to -150°C).

Important Dates in the Conquest of the Moon

2283 B.C. Mesopotamian scrolls mention sightings of a lunar eclipse.

CIRCA 459 B.C. The Greek philosopher Anaxagoras notices that the light of the Moon comes from the Sun and thus explains eclipses.

CIRCA 355 B.C. The Greek philosopher Aristotle cites lunar eclipses to support his theory that the Earth is round.

CIRCA 280 B.C. The Greek astronomer Aristarchus is the first to suggest that the Earth turns on its own axis and revolves around the Sun. He tries to measure the distances of the Earth to the Moon and Sun, but without success.

CIRCA 150 B.C. The Greek astronomer Hipparchus measures the amount of time it takes the Moon to revolve around the Sun.

CIRCA 74 B.C. The Greek philosopher Posidonius explains the effect of the Moon and Sun on tides.

PAGES 14-15
End of the World, c. 1900.
COLLECTION ANATOLE JAKOWSKY, PARIS.

OPPOSITE
Donato Creti (1671-1749).
Moon: Astronomical Observation. VATICAN MUSEUMS, VATICAN CITY.

CIRCA A.D. 150 The Greek astronomer, geographer, and mathematician Ptolemy discovers the irregularity of the Moon's movement in its orbit. His theories (not entirely accurate, since he assumed the Earth to be the center of the universe) were accepted as law until the Renaissance.

1543 The Polish astronomer Nicolaus Copernicus suggests that the Earth and other planets revolve around the Sun—a theory that casts serious doubt on Ptolemy's hypothesis and gives rise to widespread criticism, especially from the Church, since it denies the Earth its pivotal role in the universe. Nevertheless, modern astronomy is based on the work of Copernicus.

1609 Galileo Galilei, an Italian physicist, astronomer, and mathematician, first utilizes the telescope as an astronomical instrument, thereby discovering (in 1610) the uneven nature of the surface of the Moon. Until then, only intriguing dark spots had been detected, inspiring many legends. An adherent

> One midnight, with the moon
> I was going toward a road, asking her kindness, close
> to the garden of flowers, I became.
>
> Jalāl ad-Dīn ar-Rūmī, "Hand in Hand with the Moon"

Galileo Galilei (1564–1642). *Observations of the Moon.* Pen and brown ink on paper. BIBLIOTECA NAZIONALE, FLORENCE, ITALY.

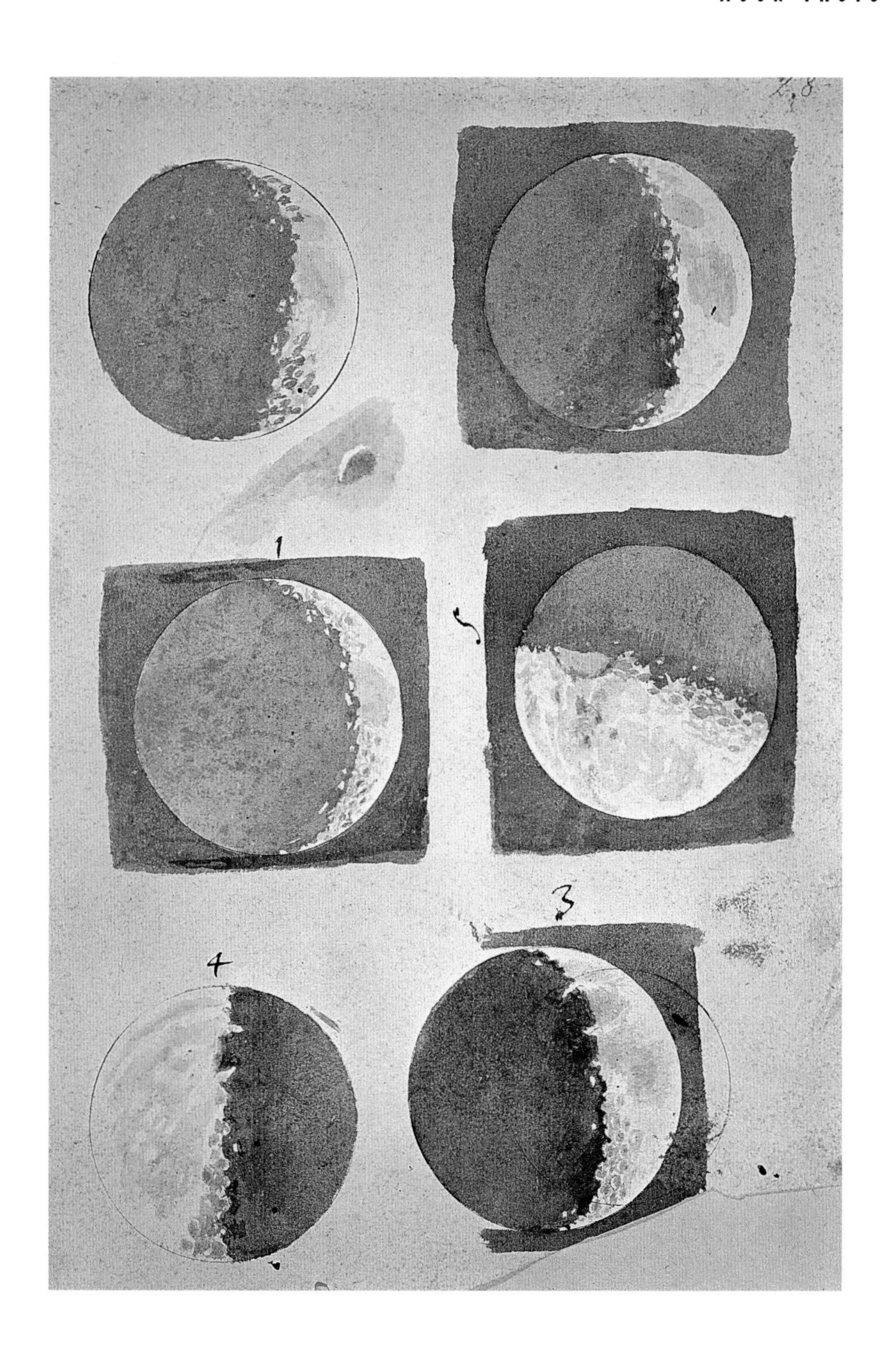
1
3
4

Earth, Moon, and Sun.
Italian, 18th century. Mixed media.
UNIVERSITY OF BOLOGNA, ITALY.

of the Copernican theory of the universe, Galileo was to have serious problems during the Inquisition.

1665 Isaac Newton, an English physicist, mathematician, and astronomer, discovers the law of gravitation thanks to the Moon, when he wonders if the Moon's rotation around the Earth and the free fall of an object (his famous apple) might not possibly be due to the same cause.

1850 William C. Bond and J. A. Whipple, of the Harvard Observatory, photograph the Moon.

1926 Robert Goddard, a college professor in Massachusetts, launches the first liquid-fueled rocket—all on his own, without government support or subsidy. Despite its modest performances—rising only 39 feet (12 m) and covering only 184 feet (56 m) at a speed of 60 miles per hour (96 km/h)—his rocket represents a giant step forward in the conquest of space. Goddard, arguably the father of American space exploration, also envisages the project of a rocket landing on the Moon with an explosion visible from Earth.

1959 The Soviet Union launches *Luna 2,* the first spaceship to land on the Moon. *Luna 3* sends back to Earth the first photographs of the hidden face of the Moon.

February 3, 1966 Thanks to retro-rockets, the Soviet rocket *Luna 9* is the first to land on the Moon without a crash, when it gently touches down in the Ocean of Storms.

Red Grooms (b. 1937).
The Astronauts, 1972.
Mixed media, 156 x 360 x 180 in.
(396.24 x 914.4 x 457.2 cm).
COLLECTION OF ROBERT ABRAMS,
NEW YORK.

Andy Warhol (1928-1987). *Moonwalk,* 1987. One from a portfolio of two screenprints, printed on Lenox Museum Board, 38 x 38 in. (96.5 x 96.5 cm).

December 21, 1968 *Apollo 8* is first manned spacecraft to orbit the Moon. Frank Borman, James Lovell, and William Anders spend their Christmas Eve in a spin. This triumph signals that the United States has overtaken the Soviet Union in the race to put a man on the Moon.

July 16, 1969 *Apollo 11* takes off from Cape Kennedy with Neil Armstrong, Edwin Aldrin, and Michael Collins aboard. On July 19, it is in orbit around the Moon. On July 20, the module lands on the Moon in the Sea of Tranquillity; at 9:56 P.M. E.S.T.,

> The stars . . . require constant observation. They have to be watched over as carefully as a pan of milk on the stove. They are always on the move. Especially the Moon. One cannot let it out of one's sight for a second. It is always up to something. For instance, this week: I left it at Nice, not only eight days or so ago, thin as a wire, next to a star, the only one in the sky (but bright as a diamond); and in Paris I find her enormous, frightfully red, swollen, sick, no higher than the housetops, completely out of proportion. A real Latin-style Moon announcing the death of Caesar.
>
> I wanted to check up on this disastrous character. All I found was, at the top of the sky (proof that it had moved about), a sort of pallid, yellow, cold, middle-sized Moon. The true Moon. I simply cannot understand it.
> ALEXANDRE VIALATTE, *Et C'est Ainsi qu'Allah est Grand*

Neil Armstrong, wearing a spacesuit weighing over forty pounds, places his foot (the left one) on the Moon's surface and starts to frog-walk, "That's one small step for a man, one giant leap for mankind," he said. We'll never know if this was a spontaneous remark or one carefully prepared by NASA's public relations department–which would be disappointing. In any case, Armstrong and Aldrin make the most of their visit to the Moon, gathering specimens, taking photographs, and exploring their immediate surroundings. Between 1969 and 1972, the Americans bring back almost 880 pounds (400 kilos) of specimens from the Moon.

1970 The Soviet Union's *Luna 16* is the first uninhabited spaceship to collect samples of lunar soil.

1972 Eugene Cernan, captain of the *Apollo 17* mission, hoists the American flag on the Moon. His is the most recent Moon landing.

Mini-Guide to the Moon for Future Tourists

Visitors to the Moon will find seas, bays, cirques, craters, and mountains; but no beaches, trees, flowers, or animals. It is a rather naked landscape. Contrary to Jules Verne's promise in *From the Earth to the Moon,* there is not yet any regular shuttle train to the Moon; but here is a guide for those who may travel there.

SEAS: The Austral Sea, the Humboldt Sea, and the seas of Crises, Fertility, Cold, Moods, Nectar, Rains, Serenity, Tranquillity, etc.

OCEAN: Ocean of Storms

BAYS AND GULFS: The Torrid Gulf, the Central Gulf, the bays of Irises, Clouds, and Dew

LAKE: Lake of Dreams

CIRQUES: Clavius, Gauss, Hipparch, Shickard, Ptolemy

CRATERS: Copernicus, Reinhold, Jules Verne, etc.

MOUNTAINS: Alembert, Caucasus, Jura, the Soviets, Leibnitz, Altai, etc.

Early Representation of Moon Dwellers "As They Were Perfectly Discovered by a Telescope," n.d. Engraving.

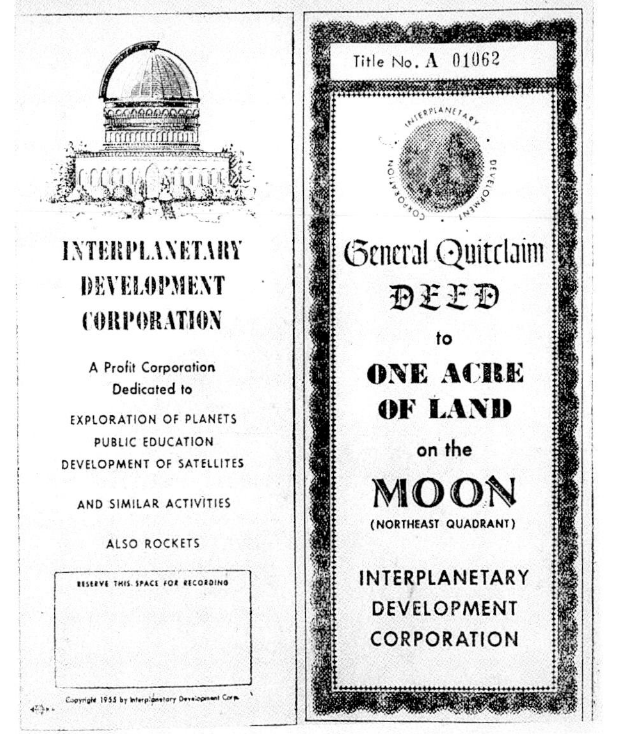

Advertisement for Land on the Moon:
$1 an Acre, 1955.

Come les yndiennes vont laver leurs enfans a la mer
quand Il est la pleine lune

Les yndiennes menent leurs enffans a la mer et
les lavent la mer et tienn ung par la main
et l'autre se mett sur son dos et les netoient bien
de facon qu'ilz n'ont aucune galle ny rongnes
et sont netz par tout le corps et nagent lesd.
yndiennes comme poissons a la mer

While the names of lunar sites are borrowed from the Earth, some earthly names have been inspired by the Moon.

- Mahābād (literally and approximately "a place haunted by the Moon") was the capital of the short-lived Republic of Iranian Kurdistan in 1946.
- Deir el-Qamar (literally, "Convent of the Moon") is a Maronite village in Lebanon, thus named because it was once the site of a convent dedicated to the Virgin Mary, in which she was portrayed standing on a crescent Moon.
- Beneath the surface of the Pacific Ocean is a range called the Moonless Mountains.
- In the state of Utah in the United States is a Moon Lake.
- In northern Argentina there is a Lago Luna (Moon Lake).
- In English, Heçn el-Qamar means "Fortress of the Moon"; Ghobbat el-Qamar means "Bay of the Moon"; and Djebel el-Qamar, "Mountain of the Moon."
- In *The Thousand and One Nights,* splendid princesses are called Badr al-Badour (the Full Moon of Full Moons), Sitt al-Bodour (Lady of the Full Moons), and Qamar az-Zamān (the Moon of the Times).

*

How the Indian Women Wash Their Children in the Sea When the Moon Is Full. From *Natural History of the Indies,* the "Drake Manuscript," late 16th century. 11½ x 7⅞ in. (29.2 x 20 cm).
THE PIERPONT MORGAN LIBRARY, NEW YORK.

School of Raphael (1483-1520). *The Creation of the Sun and the Moon,* 1518-19. Fresco. LOGGE, VATICAN PALACE.

*

- In China, the Han dynasty emperor Wu Ti, who reigned from 140 to 87 B.C., had a terrace built in his palace park, the *fou-yue t'ai,* which means "for viewing the Moon from below." But for the Moon to be viewed "from below," it had to be reflected in water. So the emperor added a lake at the foot of the terrace–the Lake Reflecting Divine Maidens, so-called because women disguised as goddesses would sail and cavort on it in the moonlight. Their lake boats, which could hold hundreds of people, were given such names as *Moon Walker* or *Moon Crash.*

- The main entrance to the private Ming dynasty garden of Zhongshan, on the south side, is called "the Moon Gate." When looking northward through it, only certain portions of the garden are visible, divided into areas of light and shade–complementary manifestations of yin and yang. When the visitor moves his eyes from the white wall to the dark vestibule, or from the sunlit garden to the shadowy Min Room with its luminous windows in the background, the contrasts create a feeling of rhythm, distance, and space.

- In Ontario, Canada, is a town called Moonbeam.
- In Austria there is a Mondsee (Moon Lake) and a mountain named Spitze Mondschein (Moonshine Peak).

Originally, the Moon and the Moon God Were One and the Same

The earliest documentary reference to the Moon appeared around 2600 B.C. in the form of a Sumerian ideogram. Although there was no specific symbol for it, we can suppose that the hieroglyphic for *luminary* designated the Moon as well as the Sun. On the other hand, there already existed a number of spoken names for the Moon god: Nanna, Sin, Ashimbar, and the Barge (the popular term inspired by the planet's crescent shape, which is almost horizontal at that latitude). The Barge is a kind of pleasure boat that carries the gods from one end of the sky to the other during the night.

The development of written word systems led to inevitable confusion between *Moon* and *Moon god* among the cultures that adhered to the idea of a lunar month. In Babylon,

*

ABOVE AND OPPOSITE
Lin Shaozhong (b. 1924), *Moon Gate in Zhongshan Park (Former Altar of Land and Grain Built in 1421), Beijing, China,* 1998. Photographs.

the Moon was "Master of the Month," and a document records the wish: "May the gods accord me a life renewed each month, like that of the Moon." The Hebrew word *vérakh* means "month," and *yareakh* means "Moon." In Chinese, the word for Moon *(yeu)* also means "lunar month." In French as well: "It took three Moons to outfit the elephants," wrote Flaubert. The Latin *mensis* and the Greek *mene* both mean "Moon" as well as "month."

In Persian, the word for both Moon and month is *mah,* and for "fish," *māhi.* Since *māh* also expresses the beauty of one's beloved, Persian poets indulged in sophisticated wordplay. In the fourteenth century, Imad-i Faqin wrote: "Seeing thy beloved face reflected in clear waters, my heart was greatly troubled and cried out: *'Māhi!'*"–the word *māhi* meaning "You are a fish!" but at the same time expressing admiration for the beauty of his loved one's face reflected in the water like the Moon.

The Ever-Changing Aspects of the Moon

Furthermore, the Moon often changes its form and face, because when it is new, it is shaped like an arc; and after VIII days, it seems to be cut in two; and when it is full, it is perfectly round.

Bartholomaeus Anglicus

The Romans employed numerical terms to designate the different phases of the Moon: *prima luna, quarta luna.* Or else a more evocative nomenclature:

○ *NOVA LUNA* or *interlunimium* for the new Moon. Or *extrema luna,* because the new Moon marks both the beginning and the end of its cycle.

● *PLENA LUNA,* for full Moon.

◐◑ *LUNA DIMIDIA* for the first and last quarters.

☽ *LUNA CAVA* (Pliny), *luna bicornia* (Suetonius), luna corniculata (Apuleius) for the crescent Moon.

☾ *LUNA PROTUMIDA* (Apuleius) for the almost-full Moon.

⊙ *LUNA SITIENTE* (thirsty Moon), a term employed only by Pliny to denote the "invisible Moon"–the new Moon. Others called it the *luna silenti* (silent Moon).

LUNA DEFICIT (the absent Moon): an expression used to describe an eclipse, which was considered a sort of lunar fainting fit.

In addition to which, *LUNA DIES* (day of the Moon) is the origin of the word for Monday in many languages.

For the Arabs, the moon was the paramount star in the sky. They called the Sun and Moon "the two Moons" *(alqamarāni). Qamar* (Moon), *badr* (full Moon), and *hilāl* (crescent) are masculine; *chams* (Sun) is feminine. The full Moon is called *bāhir* (the Moon that surpasses the stars in brilliance and eclipses them); or else *badr,* because it runs *(bādara)* toward the Sun. The dark portion of the Moon is *mahw* (absence of light) or *shāma* (a beauty spot or mole, which in Spanish is a *lunar*).

Fully aware, and with regard to nothing, you came to visit me.
Is someone here? I asked.
The moon. *The full moon is inside your heart.*

Jalāl ad-Dīn ar-Rūmī, "Be Melting Snow"

PAGES 36-37
Elliott Erwitt (b. 1928).
Moon and Building, Avila, Spain, n.d.
Color transparency.

*

IN THE BIBLE, the Moon is mentioned in the very first chapter of Genesis. "And God said, Let there be lights in the firmament of the heaven . . . to give light upon the earth: and it was so. And God made two great lights; the greater light to rule the day, and the lesser light to rule the night . . ." (1:14-16).

By referring to the planets as "lights," the Bible stresses their function of merely shedding light, in contrast to their deification by other religions. In the Old Testament, lunar cults are considered idolatrous, therefore heretical. Nevertheless, ever since A.D. 325 the date of Easter, a movable feast, has depended on the time of the first full Moon in spring, and it is highly probable that this Christian holiday is a version of an ancient pagan festival.

FOR ROMAN CATHOLICS, the Virgin Mary (sometimes referred to as "Morning Star" or "Heaven's Gate") has all the attributes of the Moon goddess of antiquity. According to certain scholars, she also represents the three aspects of the Grecian Moon: the brilliance of Selene (luminous, full Moon), the virginity of Artemis (crescent or decrescent Moon), and the mysterious powers of Hecate (dark and absent Moon) over the invisible world.

IN JAPAN, the Moon god is called either *Tsuku-yo-mi-no-mikoto* or *Tsuku-yomi-no-mikoto*—a real brainteaser. *Tsuku* is certainly an archaic form of *tsuke* (Moon). *Mikoto* ("illustrious being") is a term attributed to gods as well as to emperors. *Yomi* may mean "that which measures time"; but if it is divided into two separate words—*yo,* meaning "the night," and *mi,* the root of a verb meaning "to see"—the result is "that which sees the night." The Japanese language has no genders, so there's no way of telling whether the lunar deity is masculine or feminine.

*

OPPOSITE AND ABOVE
Pierrot, Victim of the Moon
(playing cards), c. 1900.
COLLECTION ANATOLE JAKOWSKY, PARIS.

Logo for Moon Pie,® The Original Marshmallow Sandwich.

Moon Idioms

- over the Moon: in ecstasy, in seventh heaven.
- Moon-faced: chubby-cheeked.
- moonlighting: holding two jobs at the same time, or holding a second, clandestine, job in addition to a regular one.
- to do a moonlight flit: to move out of a home in secret, often leaving the rent unpaid.
- moonshiner: illegal liquor distiller.
- moonstruck: absentminded, lost in fantasy.
- moon-eyed: wide-eyed from fright or surprise.
- when two full Moons fall within the same month, the second one is called a blue Moon. This phenomenon can occur only when the first full Moon appears on the first or second day of the month. Considering the fact that twenty-nine-and-a-half days have to separate the second full moon from the first one, the month of February can never have a blue moon. It is obviously a rare event, rather like the famous "green flash"–whence the expression "once in a blue moon."

The Moon in Films

- *A Trip to the Moon* (George Mêliês, 1902)
- *By Rocket to the Moon* (Fritz Lang, 1929)
- *Moon over Miami* (Walter Lang, 1941)
- *The Moon and Sixpence* (Albert Lewin, 1942)
- *Once upon a Honeymoon* (Leo McCarey, 1942)
- *Moonrise* (Frank Borzage, 1949)
- *The Moon Is Blue* (Otto Preminger, 1953)
- *Moonfleet* (Fritz Lang, 1955)
- *The Honeymoon Killers* (Leonard Kastle, 1970)
- *Luna* (Bernardo Bertolucci, 1979)
- *Moonlighting* (Jerzy Skolimowski, 1982)
- *The Moon in the Gutter* (Jean-Jacques Beneix, 1983)
- *Full Moon in Paris* (Eric Rohmer, 1984)
- *Moonstruck* (Norman Jewison, 1987)
- *Honeymoon in Vegas* (Andrew Bergman, 1992)

*

Tim Burton.
The Nightmare before Christmas, 1993.
Film still.

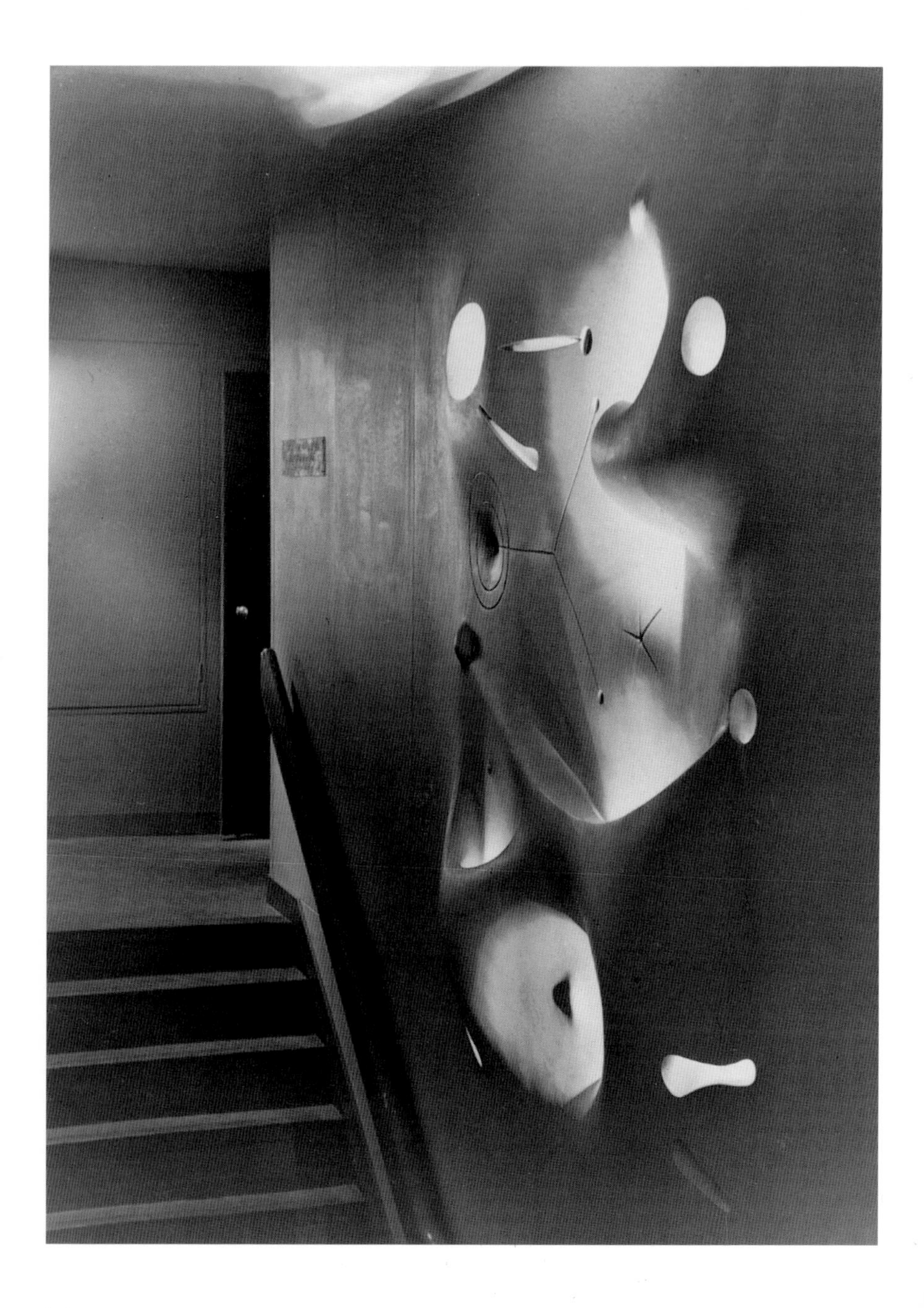

The Moon in Music

Furthermore, the Moon in the melody of the sky makes a deep, loud sound in reply to the sound of the firmament, which is shrill, if Marcien is right.

BARTHOLOMAEUS ANGLICUS

- *Everyone Feels the Joys of Love/Monostatos's Aria* (from Mozart's *The Magic Flute*)
- *Evening Sensations* (a poem by Joachim Heinrich Campe set to music by Mozart)
- *Moonlight Sonata* (Beethoven)
- *To the Moon* (a poem by Goethe twice set to music by Schubert)
- *The Wanderer in the World* (a poem by Johann-Gabriel Seidl set to music by Schubert)
- *Song to the Moon* (from Dvořák's *Rusalka*)
- *The Excursions of Mr. Broucek to the Moon and to the Fifteenth Century* (Janáček)
- *Clair de Lune* (Debussy)

*

Isamu Noguchi (1904–1988).
Lunar Voyage (sculpture for stairwell of the S.S. *Argentina*), 1948 (destroyed).

The Moon in Literature

- *True Stories Are False Fairytales* (Lucian of Samosata [b. 120 B.C.])
- *Orlando Furioso* (Ludovico Ariosto, c. 1502)
- *The Man in the Moon, or a Discourse of a Voyage Thither* (Francis Goldwin, 1638)
- *The Other World, or A Comical History of the Nations and Empires of the Moon* (Cyrano de Bergerac, 1656)
- *The Adventure of Hans Pfaall* (Edgar Allan Poe, 1839)
- *From the Earth to the Moon* (Jules Verne, 1865)
- *The Moonstone* (Wilkie Collins, 1868)
- *First Men in the Moon* (H. G. Wells, 1901)
- *The Moon and Sixpence* (W. Somerset Maugham, 1919)
- *Objective Moon* (Hergé, 1953)
- *They Walked on the Moon* (Hergé, 1954)

*

Georgia O'Keeffe (1887–1986). *New York with Moon,* 1925.
Oil on canvas, 48 x 30 3/8 in. (122 x 77 cm).
THYSSEN-BORNEMISZA COLLECTION, SWITZERLAND.

A Young Couple Dancing, n.d.
Photograph.

The Moon in Popular Song
(by Artist)

- "Bad Moon Rising" (Creedence Clearwater Revival)
- "Blue Moon" (The Marcels)
- "Blue Moon of Kentucky" (Elvis Presley)
- "Cajun Moon" (J. J. Cale)
- "The Dark Side of the Moon" (Pink Floyd)
- "Harvest Moon" (Neil Young)
- "It's Only a Paper Moon" (Frank Sinatra)
- "Man on the Moon" (R.E.M.)
- "Mr. Moonlight" (The Beatles)
- "Moonage Daydream" (David Bowie)
- "Moonchild" (Keith Jarrett)
- "Moondance" (Van Morrison)
- "Moonlight in Vermont" (Frank Sinatra)
- "Moonlight Mile" (The Rolling Stones)
- "Moonlight Serenade" (Glenn Miller)
- "Moon on Bourbon Street" (Sting)
- "Moon River" (Andy Williams)
- "Moonshadow" (Cat Stevens)
- "Walking on the Moon" (The Police)

In Nature

The Influence of the Moon on Plants, Animals, and People

Furthermore, the Moon attracts water from the sea, as a magnet attracts iron. That is why we see the ocean swell and subside according to the path of the Moon.

Bartholomaeus Anglicus

While the Moon, with the aid of the Sun, is able to influence ocean tides, it also exerts an influence on the weather, on human beings, on all creatures living on land or in the sea, as well as on birds and plants. As Apuleius wrote, "The majestic goddess reigns supreme; human life is entirely governed by her providence. Not only domesticated animals and wild beasts, but also inanimate objects are invigorated by the divine influence of the Moon's protective light and energy. Every living creature on earth, in the sky and sea, benefits from its gains and meekly accompanies it in its losses." ☾ Ever since men began to till the land, farmers have taken into account the influence of the Moon in determining the time to plant, sow, prune, graft, and harvest. "The Moon itself has indicated the most propitious days for various labors. . . . The seventh day is favorable for planting grapevines and for taming wild oxen," Virgil wrote in his *Georgics.*

This tradition was long considered to be no more than charming folklore, until recent scientific experiments confirmed the effect of lunar rhythms on the evolution of life. In Tuscany, Italy, today, the farmer's bible is still the Bacelli lunar calendar, which was based on many of the discoveries made by the ancient Romans concerning the influence of the Moon on natural phenomena.

Which Tasks Should Be Done When the Moon Is Ascendant or Descendant, Waxing or Waning?

Often assumed to be the same, the rising (ascendant) Moon and the crescent (waxing) Moon are actually quite different. The Moon may be both ascendant and waning at the same time (or descendant and waxing), since it adheres to two different cycles: the synodic revolution and the periodic one.

*

PAGE 50
George Robbins (b. 1933).
Winter Moon Rising over Chugach Mountains, Anchorage, Alaska, 1982. Color transparency.

OPPOSITE
Rapecchiano.
Madonna and Two Saints (St. Francis of Assisi and St. Catherine of Alexandria), 1627. Oil on panel. SPELLO, ITALY.

They dined on mince, and slices of quince,
Which they ate with a runcible spoon;
And hand in hand, on the edge of the sand,
They danced by the light of the moon,
The moon, the moon,
They danced by the light of the moon.
EDWARD LEAR, "The Owl and the Pussycat"

From the new Moon to the full Moon, the Moon waxes, and from the full Moon to the new Moon it wanes. This is the synodic revolution completed in 29 days, 12 hours, and 44 minutes. The periodic revolution consists of the passage of the Moon in front of the constellations of the zodiac. It is less visible than the synodic and the full passage is completed in 27 days, 7 hours, and 43 minutes.

In the Northern Hemisphere, between December 21 and June 21, the Sun ascends from the constellation of Sagittarius into that of Gemini. Sap rises with it and nature awakens. When the Sun then descends from Gemini to Sagittarius, so does the sap and nature become dormant. The Sun completes this trajectory in one year. The Moon, however, follows exactly the same path in merely one month. From Sagittarius to Gemini, it waxes, sap rises; it is lunar springtime, time to sow seeds. From Gemini to Sagittarius, it wanes; this is lunar autumn, the time to prune, to fertilize the soil and mow the lawn if you want rapid growth.

To complicate everything, the synodic and periodic cycles, which can be combined or delayed, are obviously not coordinated. For example, in 1997, the Moon was announced as descendant from April 22 (full Moon) to May 6 (new Moon), but also as waxing from April 27 to May 10. In order to make any sense of this, it is indispensable to consult a lunar calendar.

When one examines the subject in detail, it is evident that traditions vary and even conflict from one era to another, one region to another, even among different families! But generally speaking, one can say that it is advisable to plant, sow, and graft during the waxing Moon, and gather fruit, cut flowers, prune trees, uproot weeds, pick grapes, and reap crops during the waning Moon.

As for contradictions, Pliny the Elder, in the first century, pointed out a very interesting one, still in vogue, concerning profitability and quality. It seems that if quality is the objective, it is advisable to reap when the Moon is waning, although this will delay the development of new growth generated during its waxing phase. Should quality be sacrificed in the interest of profit? The Roman farmer-author, although perplexed by the dilemma, seems to have opted for quality.

In any case, it is wise to avoid all gardening activity during the two days of the cycle when the Moon is at its *apogee* (the point of its orbit farthest away from the Earth) and at its *perigee* (the point of its orbit nearest to the Earth)–both of which are indicated in all reliable lunar calendars. Seeds planted during the lunar apogee will be slow to germinate, and those planted at its perigee, when the Moon is moving at maximum speed and exerts excessive attraction, will be prone to disease.

Paul Klee (1879-1940).
Landscape of the Past, 1918.
Watercolor on paper,
8 7/8 x 10 1/4 in. (22.5 x 26 cm).
STAATSGALERIE MODERNER
KUNST, MUNICH.

Trees

If you want a tree to be vigorous and healthy, you should plant it during the rising, waxing Moon, a few days after the first quarter, preferably at the end of the day or during the night.

Moreover, since sap descends toward the roots during the descendant phase of the Moon, and rises during the Moon's ascendance, a sickly fruit tree should be pruned while the Moon is descending, in order to gain strength from the subsequent ascendant Moon.

Wood should most definitely be cut during a waning Moon, preferably seven days after the full Moon. Pliny agreed with this; further, during the Middle Ages carpenters used only this kind of lumber, the custom even being enacted into law in some French towns. In 1500, in fact, a petition was addressed to the lieutenant general of the Languedoc, denouncing the coopers who were devastating the Cévennes Forest for wine barrels and vats. The plaintiffs' outrage, however, was caused less by the quantity of timber cut than by the time they chose for cutting it. "In the interest of maximum profit," the coopers did it "during the improper phase of the Moon," although they knew perfectly well that this would debilitate the forest.

Needless to add, now that lumberyards are obliged to provide for their customers' needs all year long, such respect for the laws of nature no longer exists. Yet we can still see evidence of their validity in the few remaining facades of medieval half-timbered houses, and the beamed ceilings of ancient buildings that have remained intact for centuries.

Chinese Legends Depicted in Ideograms as Lunar Trees, or Melons

The Chinese believed there was a tree on the Moon: a cassia tree, a kind of evergreen, "the supreme medicinal plant," according to a Han dynasty scroll. The cassia flowered during the eighth month of the year, producing little white balls rather like miniature moons. According to legend, the seed of the Moon's cassia sometimes fell to Earth on the fifteenth day of the eighth month, which was the very day dedicated to the Moon.

This eighth month is called "the month of the cassia tree"; and the phrase "to cut a cassia branch in the toad's palace" means "to qualify for the examination entitling one to be a candidate"–an examination held every three years in provincial capital cities during the eighth month of the year.

According to a Buddhist theory dating from the eighth century, there is a tree in India called the jambû on the southern face of Mount Sumeru (the central pillar of the world); and the shadow of the jambû tree is projected onto the Moon when the Moon passes behind it.

Then there is the legend of the great cucumber: One day, a kindly man came upon a wounded bird and saved its life. To

*

Moonlit Riverbank with Exotic Birds.
Ming dynasty, 1600.
Interior of painted lacquer box, length: 28 in. (71.1 cm).
VICTORIA AND ALBERT MUSEUM, LONDON.

thank him, the bird gave him a seed to plant. And from the seed there grew a plant bearing cucumbers that were filled with gold and silver. A neighbor, observing this windfall of riches, decided to save a bird from dying too. But since he couldn't find an injured bird, he wounded one himself in order to nurse it back to health. Even so, the bird offered him its thanks as well as a seed. But this time, the cucumber plant it engendered grew so tall that it reached the Moon. The man then decided to climb up to the Moon on it to seek his reward of gold and silver. But when he reached the Moon, the plant suddenly vanished, leaving him stranded there forever.

Vineyards

The Roman author Columellius wrote in his treatise on agronomy: "Do your planting between the new Moon and the tenth day, or between the twentieth and the thirtieth days. This is the best time for planting grapevines." And, he added, "White grapes should be gathered during the waning Moon, in dry fair weather, and after the fifth hour of the day, for they will then be very tasty, large and plump."

For the ancient Romans (as for us today), wine played a role in religious rites, and they took loving care of their vineyards. According to the poet Horace, who was a friend of Virgil and Maecenus, "If you raise and turn your hands towards the sky on the night of the new Moon, your fertile vines will not have the scent of that poisonous Africus" (probably a reference to some horrible ancient Roman wine).

Olive Trees

Since olive trees are at their best during the days preceding the full Moon, it is advisable to harvest their fruit a few days later, if you wish to obtain a very aromatic olive oil.

According to Varro (116–27 B.C.), the Roman author of a treatise on rural economics, a good olive oil should then be processed as follows: place it in earthenware urns for two weeks, then "blow on it to transfer the surface into other urns, and repeat the procedure every two weeks for six months, bearing in mind that it is preferable for the final step to take place when the Moon is waning."

Flowers

Lavender should be planted during the first quarter of the Moon (when it is in the sign of Libra) for its perfume to be most intense.

- Tulips increase twofold when planted three days after the full Moon.

- As a general rule, flowers should be cut during the descendant phase of the Moon: the sap descends as well, and the weakened stems are easily cut.

- Jasmine is the flower of nocturnal mystery. The souls of the Moon are said to inhabit its white blossoms and illuminate them during the full Moon.

Water

tide: the alternate rising and falling of the surface of the ocean and of water bodies (as gulfs and bays) connected with the ocean that occurs usually twice a day and is caused by the gravitational attraction of the sun and moon occurring unequally on different parts of the earth.

Merriam-Webster's Collegiate Dictionary

The Sun thus really does seem to "have a rendezvous with the Moon," as in the popular song of Charles Trenet. During the full Moon and the new Moon, the attractions of the Moon and Sun reinforce each other, causing spring tides. During the first and third quarters, their forces are in opposition and result in neap tides. The magnitude of the tide is extremely variable from one region to another, depending not only on the position of the Sun and Moon in relation to the Earth, but also on the different forms of coastline and on water depth. In the Mediterranean, for example, the tidemark is very slight, whereas it rises as high as 64 feet (19.6 m) in Nova Scotia's Bay of Fundy in Canada, and 53 feet (16.1 m) in the Bay of Mont-Saint-Michel in Manche, France.

Taking into account the relative masses of the Moon and Sun and their distances from the Earth, the disturbing effect of the Moon on a particular point of the Earth's surface is almost twice as strong as that of the Sun. In other words, despite its proud airs, the Sun plays a less important role than does the Moon where tidal movements are concerned. However, the phenomenon of tides has never been entirely elucidated, and it is quite possible that other planets also exert an influence on tidal fluctuations.

While Roman writers mentioned tides, they do not seem to have understood exactly how tides function. For example, Caesar recounted how his soldiers discovered the phenomenon during a battle and were, in their ignorance, completely taken by surprise. "As luck would have it, that very night there was a full Moon, when the ocean tide is at its height, and our men knew nothing of it."

In his *Historia naturalis,* Pliny studied the influence of the Moon on rivers. "The Nile seems to obey the Moon," he wrote, "because it begins to flood at the new Moon following the summer solstice." He also observed its influence on melting ice.

Weather Forecasts

Furthermore, the Moon indicates changes in the weather, if Bede is right in saying that when it is red as gold at the beginning, this is a sign of wind; when there are black spots on the top point of the crescent Moon, this is a sign of rain; and when it is black in the middle, this is a sign that the weather will be fine when the Moon is full. And when the Moon sparkles on the oars of those who go to sea at night, it is a sign that a storm is brewing, if Bede is right.

BARTHOLOMAEUS ANGLICUS

PAGES 64-65
J. M. W. Turner (1775-1851). *Fishermen at Sea,* c. 1796. Oil on canvas, 36 x 48 1/8 in. (91.4 x 122.2 cm). TATE GALLERY, LONDON.

El Greco (1541-1614). *View of Toledo,* c. 1600.
Oil on canvas, 47 3/4 x 42 3/4 in. (121.3 x 106.8 cm).
METROPOLITAN MUSEUM OF ART, NEW YORK.

Bartholomaeus notwithstanding, the Roman scholar Varro stated, "If on the fourth day, the points of the crescent Moon are level, it portends a violent storm at sea–unless the Moon is surrounded by a halo, and the halo is very clear, because this means that there will be no bad weather before the next full Moon." It is all rather complicated.

According to Lyall Watson, the author of *Supernature,* the Moon exerts as strong an influence on air as on water, producing "atmospheric tides." And his conclusion, after years of observation, is that heavy rains tend to occur during the days that follow a full Moon–an empirical deduction, but often confirmed by personal experience.

It has long been a traditional belief in the Western world that the new Moon brings changes in the weather. Recently, however, a new culprit has been found: "With that atomic bomb, they've certainly managed to cause havoc with the weather!" And now nobody knows what to believe. "With their meteorology and all that pollution, we don't have normal seasons anymore, that's for sure."

Nevertheless, Bartholomaeus Anglicus, back in the Middle Ages, had something to say about the role the Moon can play in combating pollution–which gives one food for thought: "Furthermore, as Albumasar says, the Moon purifies

*

OVERLEAF
Paul Fusco (b. 1930).
Yosemite, California, 1972: Moon and Mountain, 1972.
Color transparency.

the air, because its fluctuations make the air subtle and fine and cleanse it; and without this, the air would be dense with fumes that rise at night and could cause great pollution."

Fish, Shellfish, and Crustaceans

The behavior of fish is related to the ocean's ebb and flow, and therefore to the changing phases of the Moon.

In order to avoid being dehydrated by the Sun at low tide, crabs seek shelter in damp rock crannies, and shellfish close hermetically to conserve their water and oxygen. Moreover, the growth of shellfish is related to the waxing and waning of the Moon: oysters open their shells to feed when the tide is rising, and close them tight at ebb tide to protect themselves from dehydration.

"The Moon must be on the wane, since oysters, along with many other things, are thin and dry," was the graceful phrase of the Roman scholar Aulus Gellius. "The Moon nourishes oysters, fills up sea-urchins, fattens mussels and cattle," was the more prosaic statement of Lyall Watson.

During the 1960s, the American biologist Frank Brown ordered a crate of Long Island oysters to be shipped to him in Illinois. Placed in a container of seawater and protected from changes of temperature and light, the oysters continued to open at the exact times when the tide was high on the shores of Long Island. By the end of two weeks, however, they began to open when the Moon was passing over the meridian line in Illinois, in other words, at the time the tide would have been high if an ocean had existed in that inland state. The clever little things had adjusted to the theoretical local tidal timetable of the Moon.

Animals

Even the smallest living creatures seem to be affected by the changing phases of the Moon. "Ants, those tiny beings who labor even during the night of the full Moon, rest when the Moon is new," wrote Pliny–who was also interested in the behavior of monkeys. "It is believed that tail-bearing monkeys are depressed during the waning Moon and happy during the new Moon." He added that all quadrupeds are afraid of eclipses–without mentioning how tailless monkeys react to them.

Scientists are understandably skeptical about such assertions, but the few who have researched the subject have made astonishing discoveries. Our oyster-loving Dr. Brown also studied rats and hamsters. He observed that rats deprived of all time references became increasingly active whenever the Moon passed above the horizon, while hamsters would abandon their normal twenty-four-hour-day biological rhythm within a few days and adopt a lunar rhythm of twenty-four hours and fifty minutes. What is the rhythm followed by a pet hamster who lives in a cage in a kitchen and spends the night endlessly turning an exercise wheel? Nobody has yet made a study of this.

Another curious fact: It seems that eye ailments in animals increase and decrease according to the corresponding

phases of the Moon. The word *lunatic,* which means "fluctuating in mood," originally referred to the periodic ophthalmia of horses, commonly known as "Moon blindness," and once believed to be connected with the fluctuating phases of the Moon. As for cats, Aulus Gellius observed that their eyes dilate and shrink along with the corresponding phases of the Moon–although it would seem more likely due to the varying intensity of light.

Birds and Eggs

On the island of Ascension between South America and Africa there is a species of bird that mates on the night of the full Moon, then flies away and returns to mate again ten full Moons later.

*

Henri Rousseau (1844-1910).
The Snake Charmer, 1907.
Oil on canvas, $66\frac{1}{2}$ x $74\frac{3}{8}$ in. (169 x 189 cm).
MUSÉE D'ORSAY, PARIS.

Marc Chagall (1887–1985). *Night in Vence,* 1952–1956.
$38\frac{1}{8}$ x 51 in. (97 x 129.5 cm). PRIVATE COLLECTION, ZURICH.

*

Hatching also obeys certain laws of nature, in this particular case contradictory: According to Varro, "Incubation should not take place until after the next new Moon. If the eggs are brooded sooner, they will hardly ever hatch." On the other hand, some modern chicken breeders claim that eggs set under the hens the day after a full Moon, thus at the beginning of its waning phase, will produce more chicks as well as healthier ones. This rule applies, of course, only to battery chickens, because free-range fowl breed and brood whenever they feel like it.

Human Beings

The various aspects of the Moon stimulate physical humors and maladies, since it would seem that the condition of lunatics is more serious during one phase of the Moon than another.

BARTHOLOMAEUS ANGLICUS

The organic matter of the human body that grows continuously like plants—our nails and hair—are subject to the influence of the Moon. Hairdressers as well as old wives tell us to cut our hair during the full Moon, preferably an ascendant one, in order to obtain faster growth and greater luster. When cut during a waning Moon, hair will grow slower and haircuts will last longer, but the effect may be less beautiful.

Not only our hair, but also our body rhythms are influenced by the Moon. Hippocrates (c. 460–c. 377 B.C.) described the effect of the Moon's passage through the zodiacal sign governing each part of the human body. For example, since Aries governs the head, one should avoid any operation on that part of the anatomy when the Moon is passing through Aries. And since the heart is ruled by Leo, one should never undergo heart surgery when the Moon is passing through that particular sector of the zodiac. Generally speaking, in fact, one should do nothing to disturb the part of the body that is governed by the sign through which the Moon is passing. A word of advice: before succumbing to the temptation of a gourmet meal, consult your astrological calendar to make sure that the Moon does not happen to be passing through the sign of Cancer, which governs the liver.

Pliny believed that the Moon also exerted an influence on the bloodstream, "which increases and diminishes according to the Moon's degree of luminosity." The ancient Romans had already noticed that hemorrhages during surgery were most profuse at the time of the full Moon, and their observation remains valid. Even today, some surgeons prefer to delay or advance an operation in order to avoid performing it during the full Moon–while also, no doubt, refraining from telling their patients the real reason for the change of date. In any case, we know that it is preferable, as far as possible, to operate during the waning phase of the Moon: the patient bleeds less, the incision heals faster, and there is less risk of complication or infection.

In 1437, Roland the Scribe (Charles the Bold's physician) and Laurent Muste, coeditors of a therapeutic lunar almanac, disagreed about the beneficial days for prescribing bleeding and laxatives because they calculated the position of the Sun by different methods. In 1528 a calendar published in Mainz, Germany, stipulated that it was preferable to bleed "a young overweight adult at the time of the waxing Moon, and a thin one when the Moon was waning; an old man during the full Moon, and an adolescent during the new Moon." In 1541, the lunar almanac of François Rabelais, who was already well

*

Lee Coombs (b. 1940).
Harvest Moon, 1985.
Color transparency.

White in the moon the long road lies,

The moon stands blank above;

White in the moon the long road lies

That leads me from my love.

A. E. Housman, "A Shropshire Lad"

Victor Hugo (1802-1885).
Memory of the Vosges, Hugo's Village, Eagle's Head, 1850.
Washed brown ink, gouache, pencil,
$13^{1}/_{4}$ x $19^{5}/_{8}$ in. (33.5 x 50 cm).
MUSÉE VICTOR HUGO,
VILLEQUIER, FRANCE.

The moon was serene and played over the wave;
At the cool casement, to the evening breeze flung wide,
Leans the sultana, and delights to watch the tide,
With surge of silvery sheen, yon sleeping islets lave.

From her hand, as it falls, vibrates the light guitar.
She listens—hark! that sound that echoes dull and low.
Is it the beat upon the Archipelago
Of some long galley's oar, from Scio bound afar?

Is it the cormorants, whose black wings, one by one,
Cut the blue wave that o'er them breaks in liquid pearls?
Is it some hovering sprite with whistling scream that hurls
Down to the deep from yon old tower a loosened stone?

Who thus disturbs the tide near the seraglio?
'Tis no dark cormorants that on the ripple float,
'Tis no dull plunge of stone—no oars of Turkish boat,
With measured beat along the water creeping slow.

'Tis heavy sacks, borne each by voiceless dusky slaves;
And could you dare to sound the depths of yon dark tide,
Something like human form would stir within its side.

The moon was serene and played over the wave.

Victor Hugo, "Moonlight"

Astarte. Seleucid period,
c. 2nd century B.C.
Alabaster statuette.
MUSÉE DU LOUVRE, PARIS.

known as a physician, recommended November 1 as a day favorable for bleeding. The date was signaled by a cloverleaf symbol, while the most favorable days for suction cups or leeches were marked with a cone, and the days for administering medicine with the symbol of a vial.

At the time, the commerce in lunar calendars designating the most propitious days for bleeding, taking medicine, baths, purges, and placing suction cups or leeches was highly profitable. So profitable, in fact, that pirate versions proliferated. In 1534 the French Faculty of Medicine complained to Parliament that numerous "ignoramuses, empirics and charlatans" were selling all sorts of trash that was seriously compromising the nation's health. But the petition was passed on to the attorney general, who passed it on to a judge, who buried it at the bottom of the pile of papers in the bottom drawer of his desk. And lunar almanacs continued to be published by ignoramuses and charlatans.

Women in Particular

Women seem to enjoy a special relationship with the Moon. In astrology, the Moon symbolizes fertility, passivity, reflection, nighttime, humidity, liquids, unpredictability, change, fantasy, periodicity, imagination, dreams, and the feminine sex.

The feminine biological cycle is of the same duration as the lunar cycle–between twenty-eight and thirty days–and the term *menstruation* derives from the Latin *mensis,* meaning month or Moon. Some Congolese and American Indian tribes still refer to menstruation as "the Moon," while the Maoris of New Zealand call it "Moon sickness."

During the Golden Age of Feminism, the coordination of menstruation and the lunar cycle inspired a mass of embarrassingly lyrical and sometimes humorless literature vaunting the privileged status of women so closely in tune with the Moon—as well as the existence of a mysterious "sistership," from which men were naturally excluded. And it didn't stop there. In her book entitled *Lunaception,* the Canadian writer Louise Lacey described the close link between the Moon and feminine fertility, and went on to recommend the natural method of contraception provided by the Moon! The problem is that the lunar method is undoubtedly no more trustworthy than was the famous Ogino-Kraus rhythm method.

The ancient Romans believed that childbirth and its accompanying suffering are also directly related to the Moon, to which they prayed for a pain-free delivery. The Assyrians, as well, believed that a rite involving the Moon god would assuage the pangs of birth. In *Myths and Rites,* Marcel Leibovici explains: "This ritual depicts a rutting bull, representing the Moon god, who deflowers the lunar cow, followed by a frightfully painful parturition. In order to comfort the cow during her ordeal, the Moon god sends two angels down from heaven to sprinkle the animal's body with waters of deliverance."

Before it became fashionable to schedule the exact date and hour of birth by provoking it artificially for the greater convenience of the mother or even of the obstetrician, midwives dreaded nights of the full Moon: not only were the expected births more numerous, there were also more unexpected ones. Statistics compiled by Dr. Walter Menaker concerning half a million births in New York City between 1948 and 1957 confirm the empirical observations: the majority of babies are born during the two days preceding and the two days following the full Moon, with a high point occurring on the night of the full Moon.

The Subconscious

The Moon also does many evil deeds, because according to Ptolemy, the Moon causes people to be fickle and unstable and rush to and fro.

Bartholomaeus Anglicus

The Moon symbolizes dreams and the subconscious, two essentially "nocturnal" realms. According to Paul Diel, a French psychoanalyst, the lunar zone of a personality comprises its darkest forces and instinctive impulses. It is the primitive element that lies dormant within us, in our dreams and fantasies, and which shapes our deepest instincts.

Be that as it may, our mental equilibrium is certainly affected by the Moon, and it seems that during the full Moon

*

OVERLEAF
Gunter Sachs (b. 1933).
The Color White, 1995. Color transparency.

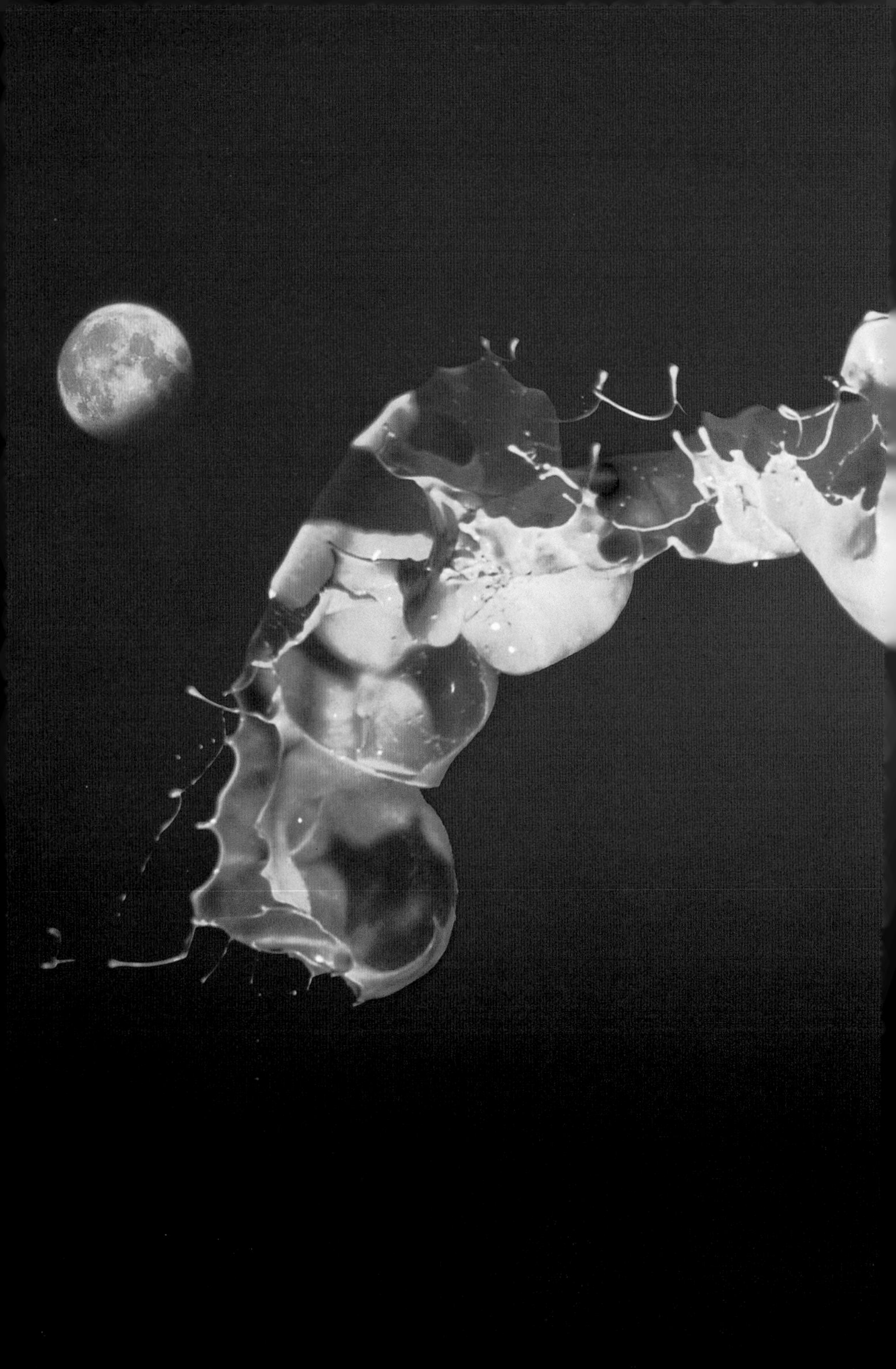

Wassily Kandinsky (1866-1944). *Red in Net,* 1927.
Oil on cardboard, 24 3/8 x 19 3/8 in. (62 x 49 cm).
YOKOHAMA MUSEUM OF ART, JAPAN.

there is an increase of crime, including rape and arson, suicide, and insanity. Epilepsy was for a long time associated with the Moon, and Babylonian writings mention people being "Moon-struck." The Greeks believed that this hallowed ailment was a form of punishment for offending Selene, and they tried to cure epileptics with mistletoe leaves gathered during the new Moon. In Palestine, it was thought that Moon stroke, a close relation of Sun stroke, caused sleepwalking. The responsibility of the Moon in this regard has never been proven, although (almost) all of us have experienced insomnia on full Moon nights. According to A. L. Lieber and Jerome Agel, the Moon's influence on the human body is transmitted by the accumulated water in our cells, tissues, and blood vessels. Its excess not only causes edema, but also increases stress. And when the attraction of the Moon (new Moon, full Moon) alters the balance of our body fluids, as happens when it also changes the tide, we undergo a psychic tension that can lead to emotional excesses.

While none of this has been scientifically proven, it has often been observed through the ages and in widely different cultures. So one must accord a certain degree of credibility to this mixture of superstition, medical science, astronomy, astrology, and empiricism.

Symbolism

In the Persian tradition, the beauty of a woman's face was likened to the beauty of the full Moon, and enamored swains often consider their beloved's radiant face more dazzling than the Moon itself. But the beauty of a face is also the reflection of a soul, and the same comparison is used in regard to virtue as to beauty: "You are like the Moon!"

When moonlight is reflected on water, a higher dimension of love is evoked: divine love. If we approach and try to touch it, the reflection blurs, then vanishes. For the mystic thirteenth-century poet Jalāl ad-Dīn ar-Rūmī, the Moon reflects the light of the Sun just as the prophet Muhammad is the reflection of Allah:

Who is it, appearing in the middle of
the night
Like moonlight?
It is the prophet of love
Come to the place of prayer.
He has brought a flame
That destroys sleep.
From the abode of the King of Kings
Of the sleepless he has come.

Mystical love is perfection. But to attain it, one must first fall in love during one's earthly existence, and the beloved is again identified with the Moon. We circle around her just as the starry sky encircles the Moon. Even folk songs were once inspired by this theme:

Thou who appears in the firmament, sublime moon,
I will turn into a star and circle around thee;
Thou who to circle around me would fain become a star,
I will become a cloud and will eclipse thee.

IN PALESTINE, according to a popular lyric:

All young women are stars, but
My beloved is the Moon!

IN CHINA, the full Moon, symbolizing the complete union of yin and yang, is compared to a mirror, an essentially feminine object. As a perfect mirror, perfectly round, it thus symbolizes perfect sexual union. A broken mirror, on the other hand, augurs the end of a love affair. Legend has it that when a husband and wife had to be separated for one reason or another,

*

PAGES 88-89
Detail of *Goddess on a Chariot: The Sun or the Moon* (see page 98).
Relief from the north gallery, Angkor Wat. MUSÉE GUIMET, PARIS.

OPPOSITE
Poster from the Theatre Palais Royal, Paris, 1934,
for the play *La Belle Isabelle.*

He who loves naturally under the sun,
adores passionately under the moon.

Guy de Maupassant, "On Water"

they would break in two a mirror bearing the image of a magpie on its back, each partner retaining one of the halves. If the wife was then unfaithful to her husband during his absence, the magpie would fly away to inform him of it.

IN THE HEBREW CABALISTIC TRADITION, the Moon is compared to an inaccessible princess who continually appears and disappears. He who woos her is fated to wander in the realm of the impossible, for to win a princess—the daughter of a king and beauty incarnate—is an unrealistic undertaking, in heaven as on earth.

THE EGYPTIANS believed that the waxing phase of the Moon was connected with the salvation of the world, while evil spirits rose when it was waning. They therefore performed a ritual especially designed to summon back the benevolent planet. In their mythology, the phrase "to bring back the full Moon" meant "to reestablish virtue in the world by restoring the wounded eye of Horus."

In all its many versions, this myth about Horus's eye is awfully complicated. To start with, Horus was the son of Isis and her brother Osiris, and also the god of the sky and stars. His birth was most unusual: After Osiris had been slain by his evil brother Seth, Isis miraculously conceived a child by her deceased husband; the child was Horus, whose destiny was to avenge the wicked murder of his father. In order to conceal the baby from the wrath of Seth, Isis gave birth to him in a marsh and brought him up in secrecy. This early part of Horus's life corresponds to the invisible phase of the Moon.

On attaining adolescence, Horus assumed his rightful place among the gods. But Seth continued to provoke him into quarrels, eventually resulting in a terrible combat, as related in the *Book of the Dead:* "The battle between the two men took place the day that Horus fought Seth, when the latter threw filth into Horus's eye and Horus tore off Seth's testicles." Although emasculation must certainly be painful, it has no effect on the Moon. On the other hand, since it was Horus's left eye that received the handful of filth, and since the left eye symbolizes the Moon (and the right eye the Sun), the Moon wanes while Horus's eye is falling from its socket and then disappears from view.

Afterward, according to the most attractive (and comprehensible) version of the story, fifteen gods, led by Thoth, the god of numbers and times, managed to put the eye back in place. The fact that there were fifteen of them symbolized the fifteen days between the new Moon and the full Moon. The waning phase of the Moon thus represents the victory of Seth and evil forces, which is accepted as predestined, with no particular ritual designed to counteract it. The restoration of the celestial eye, however, is the object of festivities on nights of the full Moon, in celebration of Horus's ultimate triumph.

A document describing the prosperity of a new reign reports: "The floodwaters are high, the days are long, the nights have their normal number of hours, and the moon arrives on time. The gods are happy."

For Muslims, the Moon (as well as the Sun) symbolizes the power of Allah. "Allah offered it to Man to help him measure

Giulio Romano (1499-1546).
Sun and Moon on Their Chariots Crossing in the Sky, 1525-35. Mural.
PALAZZO DEL TÈ, MANTUA, ITALY.

time." For desert nomads, the full Moon indicates the time when nature is at its apogee—in other words, the height of perfection. This is the most important day of the month, since all of the divine forces are assembled.

The crescent is a most significant symbol in Islam. The Fast of Ramadan, during the ninth month of the Islamic calendar, begins when the crescent moon first appears and lasts until the start of the following crescent phase. Nowadays, astronomers and astrophysicians are consulted, but in olden times astronomical calculations were considered unreliable, and faith was placed in human observation—even though this was no more infallible, if we are to believe the account of Ibn Jubayr, a North African pilgrim to Mecca in 1184: "Some of us thought they saw the crescent and pointed it out; but when they concentrated on it, it vanished, and their observations turned out to be false. . . . May God will that his crescent rise above the Muslims steadfast in their faith!"

Sun and Moon God Symbols above Figure of a God. Top of a Mesopotamian stele, usurped by the Elamites. Susa, Iran. Basalt, 24 7/8 x 15 3/4 in. (63 x 40 cm). MUSÉE DU LOUVRE, PARIS.

As early as the tenth century, the familiar crescent was extolled by the poet prince Ibn al-Mottaz as "a silver scythe that reaps narcissus blossoms from among the shining flowers of the night"—and the same image was employed by Victor Hugo several centuries later. Ever since the thirteenth century, it has figured in Muslim banners. But it was only at the end of the sixteenth century that Christians regarded the crescent moon as the symbol of Islam and the equivalent of the holy cross. And it was not until the end of the eighteenth century that it was adopted by the Ottoman Empire. Today, many Muslim nations include a crescent in the design of their national flag: Pakistan, Tunisia, and Turkey among them.

Goddess (Demeter?) with Grain and Goats. Cosmetic-box cover, Ugarit, Phoenicia, 13th century B.C. Ivory, height: 5 3/8 in. (13.7 cm). MUSÉE DU LOUVRE, PARIS.

The crescent was described in the eighteenth century by Mir Dard as a fingernail scratching the heart of Paradise in vain. And in this century Joseph Beuys, a leading figure in the German avant-garde, created an object entitled *The Moon,* which is no more than a fingernail cutting in a box.

IN BABYLON, moonless nights were interpreted as signifying the descent into hell of the dying planet. The days preceding the Moon's disappearance were therefore dedicated to the infernal deities, and it was forbidden to pray to the Moon from

Goddess on a Chariot: The Sun or the Moon.
Relief from the north gallery, Angkor Wat.
MUSÉE GUIMET, PARIS.

*

the twenty-sixth day to the twenty-ninth (the day of its abduction). Moreover, it was forbidden to embark on any major enterprise, such as a war, during the period between the new Moon and the full Moon.

THE ESKIMOS interpret the absence of the Moon more optimistically. They believe that the dead pray to it for strength to bring them back to life, and that when the Moon is absent from the sky it is accompanying the souls of the dead back to Earth. Thus, thanks to the Moon, life is constantly renewed. They also believe that the Moon exercises a direct influence on fertility: if it casts its light on a woman's naked belly, she will become pregnant.

In India, there is a similar belief: When the Moon god Soma is invisible, it is because he has descended to Earth to fertilize the waters and plants. As the *Çatapatha Brâhmana* tells it, "When during the night (of the new Moon) he can be seen neither in the east nor west, it is because the Moon is visiting the world below, entering its waters and plants."

According to another myth, the waters evaporated by the Sun are then collected in the Moon, where the gods quench their thirst. And as they drink of its waters, the Moon diminishes and shrinks, but then regenerates. The *Vishnu Purâna* relates: "It is the shining Sun that saves the Moon from exhaustion when the gods have drunk so much that only a

*

Kudurru of Nazimarutash.
Goddess Gula with Her Dog.
From the temple of Harduk, Babylon, Kassite period, 2nd half 14th century B.C. Black limestone.
MUSÉE DU LOUVRE, PARIS.

RIGHT
The Pan Painter (n.d.), Greece. *Artemis and Actaeon,* c. 470 B.C. Attic Red-Figure Krater (Ceramic), $14^{5}/8$ x $16^{5}/8$ in. (37 x 42 cm).
MUSEUM OF FINE ARTS, BOSTON; JAMES FUND AND BY SPECIAL CONTRIBUTION.

OPPOSITE
Artemis with a Doe ("Diana of Versailles"). Greece, c. 345 B.C. Parian marble, height: $78^{3}/4$ in. (199.1 cm).
MUSÉE DU LOUVRE, PARIS.

BELOW
Ergotimos. *Diana.* Detail from François Vase, 6th century B.C. Ceramic, height: 26 in. (66 cm).
MUSEO ARCHEOLOGICO, FLORENCE, ITALY.

small part of her remains. But what the Moon, Queen of the Night, loses to the gods, is restored to her by the Sun, which gathers waters from the rivers and seas for her replenishment."

And all earthly creatures benefit from this water, which is dispensed in the form of dew during nights of the full Moon.

The Mystery of Moon Spots

Before we learned of the existence of mountains and craters on the Moon, the explanation for the strange dark spots observed on its surface were imaginative and poetic. The Japanese Legend of Miyako Island is one of the most charming versions of the great mystery.

It seems that when the first settlers arrived on the island of Miyako, the Moon and Sun were lifelong friends. In a grand gesture of benevolence, they decided to offer mankind the gift of immortality. So on New Year's Eve, they sent to Earth a messenger named Akarã-zzagama, bearing a yoke on his shoulder with a bucket at either end. One of the buckets contained the water of life, destined for humanity; and the other the water of death, destined for soulless creatures such as serpents. Tired by his journey and wishing to rest for a moment, the messenger neglected his duty and laid his burden on the ground. Whereupon a serpent slithering nearby upset one of

*

Soltan-Mohammad. *Celebration of 'Id*, c. 1527.
Probably Tabriz (now Iran). From a *Divān* of Hāfez.
Opaque watercolor, ink, and gold on paper, 7¾ x 5¾ in. (20 x 15 cm).

the buckets and was inundated by the water of immortality. The messenger hastened to upturn the bucket, but it was too late: there were barely a few drops left. In despair, he then decided to pour the contents of the other bucket over mankind–and that is why men are not immortal. On his return to Paradise, the messenger recounted the mishap to the Sun, who sentenced him to stand on the Moon, bearing his yoke and barrels, for as long as men continue to inhabit the island of Miyako. And this explains the presence of those strange spots we see on the surface of the Moon.

The storyteller added a sequel to this legend: Since men thus lost their chance of immortality, the gods took pity on them and decided to at least offer them youth. And that is why "the waters of youth" are sent from heaven every year on New Year's Eve.

THE RYUKYU ISLANDS, also in Japan, have a similar legend. According to the inhabitants of Tarama, the Moon's beams

A panel from
Batman Black and White,
June 1996, Issue #1.
DC COMICS.

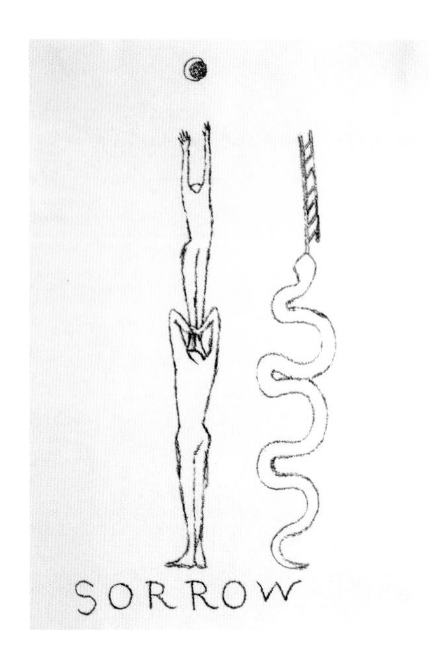

Francesco Clemente (b. 1952). *Sorrow*, 1994. Photo etching printed on handmade vellum, 11 1/8 x 8 1/8 in. (28.2 x 20.5 cm). PRIVATE COLLECTION.

outshone the rays of her husband the Sun. The jealous husband asked his wife to give him some of her light, on the pretext that night shift workers like herself do not need to see so clearly. But the Moon refused, and the Sun gave her a shove that propelled her to Earth, where she fell into a muddy field. A passing peasant, witnessing the disaster, removed his yoke and began to wash the Moon with the water from his buckets. On her return to heaven, the Moon noticed that she was not as bright and shiny as before, but still wished to reward the peasant by inviting him to visit her. And ever since then, the peasant climbs to the Moon at the time of the full Moon, and it is his shadow, with his yoke and buckets, that we can see. (But whether being stuck on the Moon with a yoke and a couple of buckets is a punishment or a reward is a matter of personal opinion.)

Variations of the same legend exist *in Siberia, Ireland, Germany, in Buddhism, and among the Tlingits* of the Pacific Northwest. In the latter's version, two little girls were carrying buckets of water, when one of them said to the other: "The Moon looks like a bone between my grandmother's lips." It was apparently the most inept remark she could have made, because the girls were immediately whisked up to the Moon, where the one who made the unflattering comment was punished by being torn to pieces, while the other survived and remains there to this day.

When the Sun and Moon live together as man and wife, their incessant quarrels cause Moon spots.

IN THE SYRIAN DESERT, the Moon is personified as a fun-loving young man, and the Sun as an old woman who runs after him. At the end of the lunar month they fulfill their conjugal duty. The result, of course, is an unending exchange of blows that leave their traces especially on the Moon.

IN TURKEY, the Sun threw mud in the face of his wife the Moon. But *the Incas* believed that it was dust cast by the Sun in a fit of jealousy because it thought the Moon outshone it. *To Peruvians,* Moon spots resemble foxes or jaguars; *to Mexicans and Guatemalans,* they seem more like hares or dogs.

The Calendar

Since these mysteries are beyond us,
let's pretend we invented them.

Jean Cocteau

Our first attempts to measure time were based on the phases of the Moon, which are shorter and more easily observed than those of the Sun. As early as the third century B.C., the Babylonians had calculated the exact time span between two lunar phases, thus defining the lunar month. Inventing a calendar involves establishing guideposts and giving ourselves the reassuring impression of being in command of things beyond us.

The lunar calendar that was observed throughout prehistoric Europe consisted of thirteen months, each twenty-eight days long (the duration of a lunar cycle), plus an extra day at the end of the thirteenth month, to compensate for the time gained during the Earth's rotation around the Sun. Total: 365 days.

The problem then was how to coordinate the lunar calendar with the solar one, which was composed of twelve months and four seasons. This seems to have been resolved during the first millennium B.C., although peasants in isolated regions continued to divide the year into thirteen twenty-eight-day months for hundreds of years.

IN PRE-ISLAMIC PERSIA, the Zoroastrian Calendar, named after the prophet Zoroaster, or Zarathustra (628–551 B.C.), divided the year into twelve months of thirty days, skipping a month every 120 years! Each month was given the name of a god; the twelfth day of every month being dedicated to the Moon goddess Makh, to whom were offered sacrifices such as: "I sacrifice to the Moon of bovine essence, goddess rich in brilliance, rich in Khvarrah, rich in clouds [when clouds appear in the sky, it is at her bidding], rich in warmth [warm], rich in splendour [wise], very beautiful, full of prosperity [it is she who endows livestock with fertility], highly gifted [for creations and decisions], rich in profit [ensuring the yield of water and plants], rich in greenery [it is the Moon who makes and keeps the lands green], benevolent [she bestows riches], healing [dispenser of all good things]" (Māh Nyâyishn, *Litany to the Moon*).

During their nomadic period, THE JEWS also based their time calculations on the phases of the Moon. In Hebrew, there are

*

PAGE 108
Detail of *Aztec Calendar,* after 1300 (see page 130).
Stone carving, diameter: 141 3/4 in. (360 cm).
MUSEO NACIONAL DE ANTROPOLOGIA, MEXICO CITY.

two different words for month: *yérakh,* which is derived from *yareakh* (Moon), and *khodesh* (the new Moon). Afterward, as settled tribes, they conserved their lunar calendar alongside the somewhat conflicting solar one.

And then the moon, ever punctual
To mark the time, an everlasting sign

ECCLESIASTICUS

If the author of Ecclesiasticus accords the Moon a preponderant role in calculating time, the partisans of the solar calendar object to certain inconsistencies as described in the *Book of Jubilees:* "For there are some people who base their observations on the Moon, although it confuses the seasons and arrives ten days early every year. They thus may turn an ill-fated day into a lucky one, an unholy day into a feast day, and every day they confuse the holy with the impure and the impure with the holy."

THE ANCIENT GREEKS, THE CHINESE, MONGOLS, AND INDIANS adhered to a lunar-solar calendar, that is to say, a solar calendar adjusted as far as possible to correspond to the lunar month. Today the Catholic church still utilizes this kind of calendar when setting the dates for movable feasts such as Easter.

*

PAGES 112-113
Caspar David Friedrich (1774-1840).
Man and Woman Watching the Moon, 1824. Oil on canvas,
13 3/8 x 17 3/8 in. (34 x 44 cm). NATIONALGALERIE, BERLIN.

Now came still evening on, and twilight gray
Had in her sober livery all things clad;
Silence accompanied, for beast and bird,
They to their grassy couch, these to their nests,
Were slunk, all but the wakeful nightingale;
She all night long her amorous descant sung;
Silence was pleas'd: now glow'd the firmament
With living sapphires: Hersperus that led
The starry host, rode brightest, till the moon,
Rising in clouded majesty, at length
Apparent queen unveil'd her peerless light,
And o'er the dark her silver mantle threw.

JOHN MILTON, *Paradise Lost*

Our present-day calendar evolved from THE ROMAN CALENDAR. "When the founder of Rome established the calendar, he decided that the year would comprise two times five months," wrote Ovid. So the earliest Roman calendar consisted of only ten months, the first of which, dedicated to Mars, was called "Martius." In 153 B.C., Romulus's successor, Numa, added two more months, Januarius and Februarius, at the beginning of the year. In the Numa calendar, there were four months of thirty-one days, seven of twenty-nine days, twenty-eight days in the month of February, with a month of either twenty-two or twenty-three days (called Mercedonius) added every other year. The Kalends (from which the word *calendar* is derived) was the first day of the month, coinciding with the new Moon. The Ides were the fifteenth day of March, May, July, and October, and the thirteenth day of the other months, all of them days of the full Moon. This lunar year consisted of 355 days.

In 45 B.C., Julius Caesar reformed the Numa calendar. The Julian calendar comprised alternating months of thirty and thirty-one days, but only twenty-nine days in February, and an extra day every four years, adding up to an average annual total of 365.25 days. In order to maintain the seasons in their normal place, Caesar also had to add two months of thirty-three and thirty-four days, resulting in a total of 442 days, plus the twenty-three days of Mercedonius. The number of days in the year 45 B.C. thus totaled 445—certainly the longest year

*

Man Ray (1890-1976).
The Moon Shines on Nias Island, 1926.
Gelatin silver print, 9 x 6½ in. (16.5 x 22.9 cm).

"Men have designated points in time in terms of dates for ages. Even Medieval illuminated manuscripts mention dates, although it is impossible to attribute them precisely, due to the continual change of calendar systems. Until 1000 A.D., the year began at Christmas. And even then men didn't know its exact date until Pope Julian I, in 336 A.D., decreed it to be December 25; and it became a 'holy day' only in 138 A.D., by order of Bishop Têlêsphore, although the date could vary from one year to the next. It is obviously a complicated affair. All the more so because, at the same time, other French provinces began their year on March 25. In 1564, Charles IX finally put an end to the confusion by ordering that the year should begin on the first of January, as it had for King Numa, Julius Caesar, and the rest of the world. Romulus and Charlemagne had begun it on the first of March. Then in 1792, the French government changed

in history. It was also, according to contemporaries, "a year of confusion." Nevertheless, the solar calendar was born, and the month of Quintilis was rebaptized "Julius" in honor of its creator, the statesman who had finally brought order out of chaos . . . if only momentarily. Since the astronomical year based on the Earth's revolution around the Sun was slightly shorter than the official year (365.24 days plus a few hours), the shortfall, insignificant as it was at the beginning, increased as time went on and amounted to ten days in the sixteenth century, causing a perceptible disparity with the traditional dates of the four seasons.

everything: the New Year would commence on the first day of the autumn equinox, September 22. Napoleon changed it again in 1806. But the Pope had already complicated the matter by removing ten days from the month of October in order to give the Earth time to catch up with the Sun, when he produced yet another revised calendar that was coordinated with the planets. As a result, the French gained ten extra days, thereby surpassing the English in age–until the English followed their example and thus became older than themselves, but for a shorter time. And the situation continues. . . . It's easy to see why we begin this month of January, 1963, in such a state of disorder."

ALEXANDRE VIALATTE,
from "Chronicles of the Winds, the Moon, the Fogs and Mammoths," in *La Montagne,* 1963

In 1582, Pope Gregory XIII corrected this gradual drift of the seasons in his reformed Gregorian calendar, still the official civil calendar in much of the world. In Italy, the day after Thursday October 4 was decreed to be Friday October 15. In France the adjustment was made by skipping directly from the ninth to the twentieth of December. Oddly enough, there

*

PAGES 118-119
Canaletto (1697-1768). *Night Festival at San Pietro di Castello,* 1756.
Oil on canvas, 38 3/8 x 51 1/2 in. (97.4 x 130.8 cm).
PRIVATE COLLECTION.

seems to have been little opposition to this change of dates. But in England, where the same reform was eventually adopted by the Protestant government only in 1752 (until then the English had preferred to disagree with the planets rather than to agree with the pope)–the reaction was violent. It should be added that the same year was chosen for changing New Year's Day from March 25 to January 1. Demonstrators in the streets of London shouted, "Give us back our three months!" In France, after wandering from the first to the twenty-fifth of March, with brief stopovers at Christmas and Easter, the New Year has begun on January 1 ever since 1564.

It would seem that when the French serenely accepted the overnight hop from the ninth to the twentieth of December, the idea of time was less dependent on the calendar than it was to be two centuries later, and that people then were indifferent to the loss or gain of ten days. A hundred years before, they hadn't even bothered to calculate their exact ages, often rounding them off to the previous or following decade, as evidenced by *Le Testament* of François Villon: "In the year of my thirtieth age"–referring to the year (1455) when there is documentary evidence that he was "more or less" twenty-six years old, which would have made him thirty-two, "more or less," when he composed the poem in 1461. In any case, when one considers the controversy caused by our present little

*

Sonja Bullaty.
Westminster Abbey, London, England, 1984.
Color transparency.

Alfred de Musset

There was, in the dusky night,

On the yellowed steeple

The moon,

Like the dot of an i.

one-hour change from wintertime to summertime, one cannot help wondering what would be the reaction of people today, with their automatic bank payments and appointment-filled agendas, if ten days were suddenly to vanish from their lives.

Moon Festivals

The Jewish Sabbath was originally associated with the feast of the full Moon. The Hebrew word *shabbāth* is derived from the word *shābath,* which means to cease; and the Sabbath was the day when the Moon ceased to wax. Later on, it became a weekly celebration and a day of rest from work.

THE MOON FESTIVAL, one of the three most important annual events in China, used to take place on the fifteenth day of the eighth month, during the full Moon of the autumnal equinox, which is exceptionally beautiful there. The night of "luna contemplation" or *chang* (a word that also means to enjoy, to admire, to love) was the occasion for a joyous fete beside and on the water, during which odes in praise of the Moon were sung. The autumn Moon and the Moon's reflection in water were favorite themes of one of the greatest Chinese poets, Li Po (701-762).

*

Andō Hiroshige (1797-1858). *Moonlit Street Scene in Edo,* 1856.
Color woodblock print, 14 x 9¾ in. (35.6 x 24.9 cm).
VICTORIA AND ALBERT MUSEUM, LONDON.

名所江戸百景

木曽海道六拾九次之内
洗馬

PAGES 124-125
Andō Hiroshige (1797-1858). *Moonlight on the River Sheba,* 1837-42.
Color woodblock print. MUSÉE GUIMET, PARIS.

*

Among the flowers, a jug of wine.
Alone, without companions, I drink from it.
I raise my cup to the shining Moon.
We are three shadows, three persons.

(The three persons are presumably the Moon, Li Po, and his shadow.) To the poet, moonlight on water represented beauty and perfection, but also elicited melancholy–and called for wine. According to legend, Li Po drank too much of it one night and drowned while attempting to seize the reflection of the Moon in the water. The image of "Li Po Trying to Grasp the Moon" was, in fact, a frequent subject of paintings during the Song dynasty.

THE TSUKI-MI FESTIVAL, imported to Japan from China around A.D. 897, was also devoted to contemplation of the Moon; it too took place on the fifteenth day of the eighth month and consisted of feasting, music, and poetry contests. This tradition was maintained by the imperial court until the thirteenth century, when it spread to the upper classes, whose members improvised *waka* (poems of 31 syllables) and *haiku* (poems of 17 syllables). From the eighteenth century, it was adopted by the bourgeoisie and country folk, and even peasants could be seen composing *haiku* as they contemplated the Moon. Today, offerings of cakes, sweet sake, rice awns, and

Adam Elsheimer (1578-1610).
The Flight into Egypt, 1609.
Oil on panel, 12 3/16 x 16 1/16 in. (31 x 41 cm).
ALTE PINAKOTHEK, MUNICH.

*

autumn flowers are still made to the Moon in gardens and on verandahs.

IN THE ALTAI MOUNTAINS OF CHINA, several calendar systems coexist. Along with the scholarly calendars borrowed from the Chinese, there is another based on atmospheric and economic phenomena: Moon of the *first frost, black weather, intense cold, middle of winter, month of the reindeer, first fish,* etc.

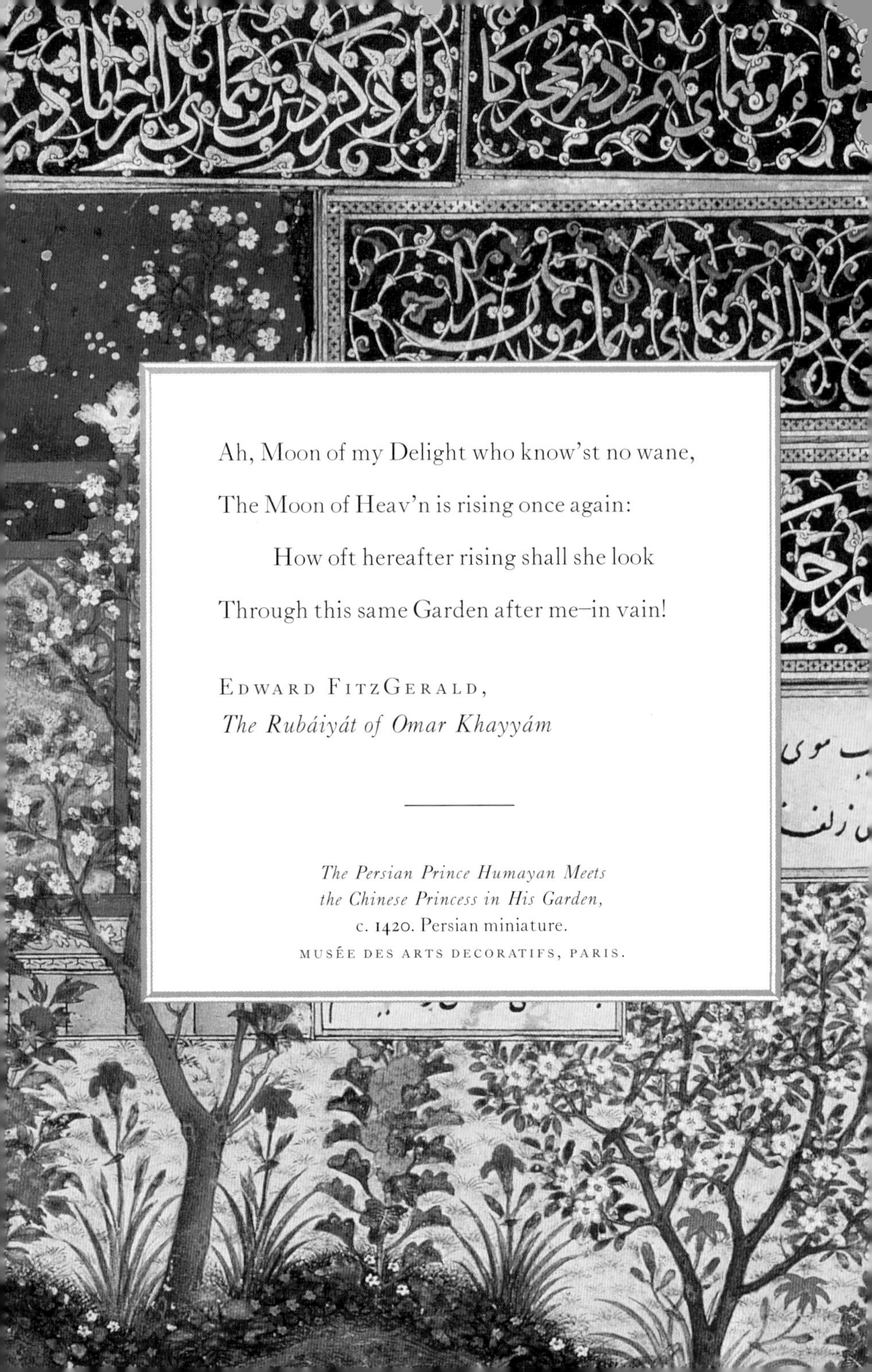

Ah, Moon of my Delight who know'st no wane,

The Moon of Heav'n is rising once again:

How oft hereafter rising shall she look

Through this same Garden after me—in vain!

Edward FitzGerald,
The Rubáiyát of Omar Khayyám

The Persian Prince Humayan Meets the Chinese Princess in His Garden, c. 1420. Persian miniature.
MUSÉE DES ARTS DECORATIFS, PARIS.

در حکمت گوید
میان تو یک موی و از موی کم
چو هندوی زلف تو بر آتشم
من از غم چو مویی و در موی خم
ز خورشید روی تو در آتشم
ز نقش رخت نسخه دیده ام
چه نقشی که مثل تو نشنیده ام

Aztec Calendar, after 1300.
Stone carving, diameter: 141¾ in. (360 cm).
MUSEO NACIONAL DE ANTROPOLOGIA,
MEXICO CITY.

*

IN SIBERIA, where a very brief summer is followed by an interminable, dark winter lit by the Moon and stars when there is no storm or fog, the idea of time is very relative and its division into months would be of little practical significance. Festivals there are scheduled when the year "turns its head" in spring and fall.

The most important religious festivals in India—those of Diwali and Dassera, Krishna's birthday, and the Night of Shiva—are movable feasts, their dates determined by the lunar calendar and their names including the day of the month. For example, Nâg-Panchâmi (the Fifth Serpent) is the festival of serpents that occurs on the fifth day of the two-week moonlit period during the month of Vaishakh; and the birth of Krishna is called Gokul Ashtamî (the Eighth of the Stable). In addition, at every full Moon and every new Moon, sacrifices are offered to the principal Vedic deities—Agni, Soma, and Indra. Around 200 B.C., when the Vedic religion gave way to Hinduism, the same rites continued to be practiced, but it was Vishnu who was worshiped.

In the literary imagery of old Persia, the Sun and Moon are personified as two sultans seated on a rug, one of them all alone and the other accompanied by a galaxy of stars; or as two birds arriving from the east, one of them magnificently feathered, and the other deplumed; or as twin sisters, one of them sterile and the other giving birth to daughters (the stars).

Sometimes the two planets represent the path of destiny and are pictured as two torches that light the way; or as two windows opening onto the garden of destiny, in which the gardener opens one window at sunrise and the other at sunset; or as two mirrors in which Lady Fate observes her image, in the morning in one, in the evening in the other; or as two rosebuds, one of which bursts into bloom at daybreak, and the other at nightfall; or else as two glasses of wine, one imbibed in the morning and the other in the evening.

The Sex of the Moon

The planet Moon is cold and excessively moist; it is female and nocturnal and resides in each sign II days and six hours, and completes its path in XVIII days.

BARTHOLOMAEUS ANGLICUS

THE MOON,
which merely reflects the light of the Sun, passes through different phases and changes shape; it symbolizes dependency, passivity, periodicity, renewal, and fertility. It is therefore generally considered to have a feminine nature. Wherever it is considered male (in Japan, Germany, and the mountains of South Vietnam, for instance), it is because the feminine aspect associated with fertility is considered to be active and thus masculine.

IN CHINA, the Sun and Moon are inanimate, sexless celestial bodies. Though one cannot say that the Sun is masculine and the Moon feminine, one can identify them as being either *yang* (luminous, dry, warm, male) or *yin* (shadowy, moist, cold, female). The Sun is obviously yang and the Moon yin.

PAGE 132
Detail of *Diana of Ephesus* (see page 139). Roman, c. 1st-3rd century.
Bronze, blackened bronze, and alabaster.
MUSEO ARCHEOLOGICO NAZIONALE, NAPLES.

*

IN CAMBODIA, traditions vary: the Moon is sometimes the husband, sometimes the wife, of the Sun.

AMONG THE AMERICAN INDIANS, Brother Moon and Sister Sun live together.

IN NORTHERN ASIA, the Sun, always benevolent in that harsh climate, is often female. The Moon is more ambiguous. Husband, brother, father, daughter, or sister, it is associated with fertility, but also with cold and death.

AMONG THE INCAS, the Moon had four faces: first as a goddess unrelated to the Sun; then as the god of women; then as the wife of the Sun; and finally as the incestuous wife of the Sun, who is her own brother. Queen of the skies, she reigns over the sea and winds, and over earthly queens and princesses as well; she is also the protectress of childbirth. In short, she is in charge of everything feminine.

*

Thot, God of the Moon. Detail of sarcophagus.
Butehamon, Egypt, 21st dynasty.
MUSEO EGIZIO, TURIN, ITALY.

PAGES 136-137
Anne Louis Girodet de Roucy-Trioson (1767-1824).
Endymion's Sleep, 1793. Oil on canvas, 78 x 102¾ in. (198 x 261 cm).
MUSÉE DU LOUVRE, PARIS.

Vaguely illumined by the summer moon
Upright, naked, dreaming in her golden pallor
Mottled by the heavy tide of her long blue hair,
In the shadowed glade where the moss is bedecked with stars,
The Dryad beholds the silent sky . . .
—White Sélène lets float her veil,
Timidly, over the feet of the fair Endymion,
And throws him a kiss in a pallid beam . . .
—The Spring cries far away in a long ecstasy . . .
It is the Nymph who dreams, an elbow upon her urn,
Of the handsome young man that her wave has entreated.
—A breeze of love has passed in the night,
And, in the sacred woods, amidst the dread of the great trees,
Majestically erect, the shadowy Marbles,
The Gods, on whose brows the Bullfinch has nested,
—The Gods listen to Men and to the infinite World.

ARTHUR RIMBAUD, "Sun and Flesh"

FOLIES-BERGÈRE
ILKA
DE MYNN
LITH. F. APPEL, 12, R. du DELTA, PARIS

AMONG SOUTHERN SEMITIC (Arabian, Saudi Arabian, Ethiopian) people the Sun is feminine, because it is a source of suffering for desert nomads; whereas the Moon, which provides the caravan with rest and guidance during the night, is accorded the superior rank of masculinity–superior, at least, in those patriarchal societies.

THE ANCIENT GRECO-ROMAN MOON was a woman with three different faces:

- Selene, the full Moon. When shepherds could be kings, mortals could wed goddesses. Endymion, the king of Elide (no insignificant little monarch but the son of Zeus/Jupiter and a nymph), married Selene and fathered fifty daughters with her. The full Moon thus seems to be associated with fecundity.

*

Diana of Ephesus. Roman, c. 1st–3rd century.
Bronze, blackened bronze, and alabaster.
MUSEO ARCHEOLOGICO NAZIONALE, NAPLES.

OPPOSITE
Poster from the Folies Bergère, Paris, 1897.

- Artemis/Diana, the ascendant or descendant Moon, is portrayed with a crescent in her hair, a bow and arrow, and hunting dogs. As a child, she asked her father, Zeus/Jupiter, among other favors, to endow her with eternal virginity. She scorned love and wished to be the equal of her brother Apollo, symbol of the Sun. The Temple of Diana in Ephesus, Turkey, was one of the Seven Wonders of the Ancient World. It was said that her statue there had fallen from heaven, because it represented the epitome of feminine perfection on Earth.

- Hecate, the new Moon, obscure and absent, reputedly evil, is sometimes pictured prowling at night in the netherworld, followed by baying hounds and sometimes sporting three heads: those of a dog, a lion, and a mare–hence her name: three-headed (or three-faced) Hecate.

The Stormy Marriage of Sun and Moon

When the Sun and Moon form a couple, their marital life is hardly serene: there is incest, enucleation, numerous disputes, infidelities, revenge, and more.

*

Jackson Pollock (1912-1956).
The Moon Woman, 1942.
Oil on canvas, 69 x 43 1/8 in. (175 x 109.5 cm).
THE PEGGY GUGGENHEIM COLLECTION, VENICE.

As the moon sidles up
Must she sidle up,
As trips the scared moon
Away must she trip;
'His light had struck me blind
Dared I stop.'

She sings as the moon sings:
'I am I, am I;
The greater grows my light
The further I fly.'
All creation shivers
With that sweet cry.

William Butler Yeats, "He and She"

PAGES 142–143
George Robbins (b. 1933).
Early Morning Sun Lights up Heavy Timber as Full Moon Sets Behind Beartooth Mountains near Red Lodge, Montana, 1992. Color transparency.

*

AMONG THE ESKIMOS, the original myth is concerned with incest: It seems that Brother Moon was in the habit of visiting his sister the Sun at night and in disguise. One night, however, she recognized him, and in a rage cut off her breasts and flung them in his face. After which they chased each other around the tent, each of them bearing a torch. The circular movement gradually spiraled them into the sky, where they remain. Sister Sun never forgave her brother and, militant feminist before her time, rejoices in every setback of the "male" element on Earth. Brother Moon, meanwhile, enjoys great prestige. He rules over hunting and influences the daily lot of men by feeding or starving them; he controls natural phenomena such as rain, snow, earthquakes, storms, and tides. But he is also the guardian of morality, and strictly enforces the observance of taboos—even though he himself lightheartedly infringed the one prohibiting incest.

AMONG THE MAYAN INDIANS, the Sun and Moon inhabited the Earth before changing into celestial bodies. They lived as a couple and quarreled endlessly. One day, the Sun even tore out his wife's eye!

IN OLD PERSIA, the Moon was a seducer (male) who forced his attentions on the Sun (female). Again, disputes ensued.

After the Sun blinded the Moon by tossing her hair in his eyes, the Moon threw needles in her face. And since then, one cannot look directly into the Sun, for its rays are those sharp needles. According to Farīd od-Dīn 'Attār, a mystical thirteenth-century poet, at the moment when the Moon is about to rejoin the Sun, she cannot bear his brilliance and so sinks into the lower world.

IN INDIA, the Sun took the Moon to be his wedded wife and she bore him a host of children (the stars). But one day the Moon became bored with conjugal life and was unfaithful to the Sun who, in a rage, cut her in two. Whereupon he regretted his deed and permitted the Moon to rise to heaven, at least for a few days, with all her splendid beauty intact.

Eclipses

The Moon deprives us of the light of the sun when she is placed between us. The moon loses its clarity when it moves into the shade of the earth, and due to the dirty air near the moon, it is soiled and uglified, or so says Marcien.

BARTHOLOMAEUS ANGLICUS

A lunar eclipse is the temporary (total or partial) disappearance of the Moon when it passes in the shadow of the Earth. A solar eclipse is the temporary disappearance of the Sun when the Moon is situated between it and the Earth.

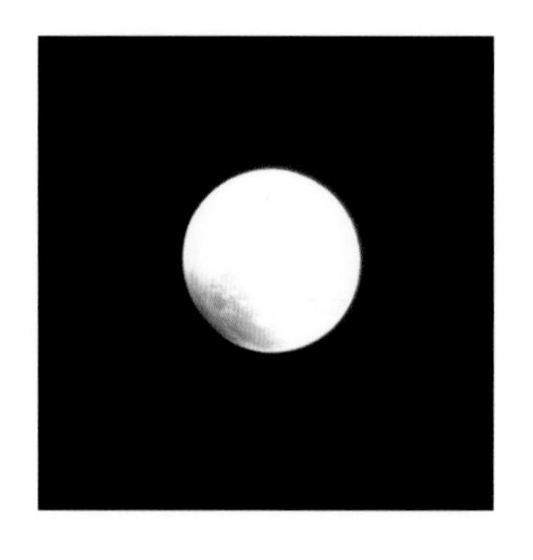

The temporary disappearance of the Sun can have strange consequences, as Woody Allen described in "Count Dracula." It is a well-known fact that broad daylight is unhealthy for vampires, who usually fall to pieces when struck by the slightest ray of sunshine. That is why Allen's Dracula rose from his coffin only after nightfall, some mysterious instinct telling him when the hostile planet had gone to bed. Once, when this instinct had thus informed him as usual, he raised the lid of his upholstered coffin, stuck his nose out, observed that it was dark, and hastened to pay a visit to the bakery, for he was attracted to its proprietress. The lady greeted him by saying, "Ah, you've come to watch the eclipse with us!" A sad ending to the story for Dracula.

Needless to say, before there was a scientific explanation for the phenomenon, an eclipse was a terrifying event. The disappearance of a planet–its apparent death–could hardly be interpreted as an auspicious omen, and so it was assumed to be a portent of disaster.

Around 2200 B.C., the Mesopotamians observed lunar eclipses for the first time and learned to predict their recurrence, although not to understand their cause nor consequently to recognize them as natural phenomena. In the fifth century B.C., the Greek philosopher Anaxagoras, realizing that the light of the Moon came from the Sun, at last explained the

mystery. (He also asserted that the basic principle of the universe was intelligence, which shows how extraordinarily optimistic he must have been.) Around 355 B.C., Aristotle utilized the observation of lunar eclipses to prove that the Earth is round: since the shadow of the Earth on the Moon is round, it can only be caused by something round.

Most Roman writers knew that when the Moon passes in the shadow of the Earth, there will be an eclipse. Educated people therefore understood the phenomenon and calmly accepted it as a natural occurrence, while those less well informed continued to be terrified. Pliny the Elder wrote, "We have long been able to calculate in advance not only the times of day and night, but also the times of solar and lunar eclipses; nevertheless, a large proportion of the population still believes that the phenomenon is caused by magic spells and herbs, and that this form of science, the only one practiced exclusively by women, is the most reliable."

The predictability of eclipses has always been a way for the enlightened to impress the superstitious. Persian priests predicted eclipses but were careful to keep their scientific knowledge secret in order to maintain power over the ignorant. Many historical events have been associated with eclipses, including the defeat of the Athenians in Sicily in 413 B.C., and the defeat of Darius by Alexander the Great in 331 B.C.; or, in 168 B.C., the

victory of the Roman general Paulus Aemilius in the battle of Pydna, thanks to the prediction of an eclipse by Sulpicius Gallus. In the fictional worlds of Tintin and of Mark Twain's Connecticut Yankee, the time-traveling heroes escape execution by "predicting" eclipses that each one remembered from his history studies.

After the discovery of America and the development of long-distance navigation, the problem of determining longitudes was solved for a while thanks to eclipses–pertinent information being drawn from the phases of the eclipse (before the invention of the telescope), or from the shadow on the craters (after the invention of the telescope). This technique was practiced by cartographers from the beginning of the seventeenth century, and enabled them to map newly discovered lands, as well as to correct some longstanding errors concerning familiar territory. For example, after the eclipse of 1634, scientists in France and in the Near East observed that the length of the Mediterranean Sea, which until then had been measured in terms of sailing routes, was inaccurate. It had to be shortened by some 625 miles (1000 km).

Divine anger

In the Bible the lunar eclipse is a manifestation of divine anger and is supposed to be one of the signs of the Last Judgment.

Sign of Allah

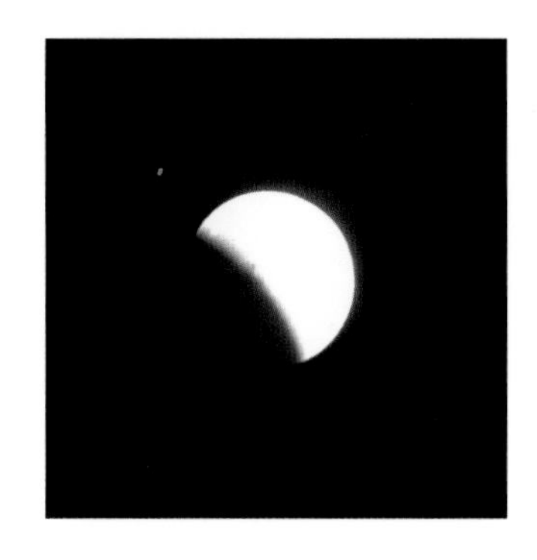

According to the Koran, too, the Day of Judgment will be announced by an eclipse. "But on Judgment Day, which is imminent when the Moon is seen to split, it will rejoin the Sun and vanish." Generally speaking, an eclipse was the herald of death—although this belief greatly irritated the Prophet, who is reported to have said: "The Sun and Moon are two of the signs of Allah. An eclipse is the cause of death to no one, but Allah utilizes eclipses to inspire awe in his servants."

Sad pallor and magic spells . . .

In Morocco, the Moon and the Sun become pale when saddened by evil events, and this explains eclipses. In Tunisia, the Moon loses its way and wanders onto the path of the Sun. Or else, she goes into hiding because a witch wishes to involve her in her magic spells.

Feminine domination

In ancient China, the tremendous disturbance represented by an eclipse was due to a trifling discord: that of the emperors and their wives. A solar eclipse symbolizes the domination of *yang* (masculine light) by *yin* (feminine darkness). One must therefore come to the aid of the Sun by shooting arrows at the monster that is devouring it.

The Incan empire may fall . . .

The Incas had four explanations for an eclipse: (1) A jaguar or a serpent is devouring the planet; (2) The planet is dying from some malady; (3) The Sun is hiding because it is angry with mankind; (4) The Moon seduced the Sun, causing the union of the two planets. In any case, the phenomenon is malevolent, and one of the premonitory signs of the end of the Incan empire will be an eclipse of the Sun.

The Moon devoured

For the ancient Egyptians, an eclipse accompanied or announced a catastrophe. The Chronicles of Osorkon, inscribed in the great temple at Karnak, describe a particular lunar eclipse: "Although the skies may not have swallowed the Moon, a great revolution occurred there." We know this was not a new Moon but an eclipsed Moon, since the event took place on the sixteenth of the month when the Moon was full. The Chronicles of Osorkon also provide formulas for averting an eclipse and magic spells to bewitch the dragon Apophis, who tries to prevent the Sun from rising every day.

Under the wings of an angel . . .

In Turkey, the Moon is hidden by fairies and *djinns* or by angels who stretch their wings to protect him from the monsters advancing on the Sun, with the intention of suckling her. Or else, it is sometimes said that the Moon disappears behind a hill, just as an ordinary hiker disappears for a moment in an undulating landscape.

Râhu the monster

Documents kept by the Commission of Customs and Traditions of Cambodia relate the story of Râhu, the monster who devoured the Moon and Sun. It seems that Râhu was born more or less as the result of a family quarrel.

Legend has it that long ago, there lived a rich merchant and his three sons. One day when the three brothers were cooking rice, the youngest made the fire too hot; the boiling water overflowed and extinguished the fire. During the altercation that followed, the eldest hit the youngest with a ladle, and the middle brother came to his defense. In the end, they all shared the rice: the portion of the eldest served in a golden bowl, that of the second son in one of silver, and that of the youngest in one of straw. When the Bodhisattva arrived, disguised as a monk, the three brothers offered him their rice, each expressing a wish: "The eldest asked to become the Sun, with golden rays that shine on the four worlds and last for nine hundred thousand years. The second-born asked to become the Moon, with its pure light clear as crystal, which shines upon the four worlds and lasts for nine hundred thousand years. As

for the youngest, he wished to be given an enormous stature and formidable strength, the very sight of which would intimidate his brothers, and also to live, like them, for nine hundred thousand years."

And that is how Râhu was born, the monster who swallows the Moon and the Sun in order to seek vengeance or else to seek pardon, depending on his mood. And the Cambodian tradition interprets eclipses (described as "Râhu catching the Moon") in a favorable or unfavorable way according to how it appears in the sky. "If the Moon enters the mouth of Râhu from the east and emerges from his body from the west, it is a sign of good fortune for the nation." However, "if Râhu swallows only half of the Moon and then emerges sidewise, there will be famine in the land."

At any rate, it is advisable to refrain from getting married, building a house, or embarking on a long voyage, during a lunar eclipse. A pregnant woman should not stay at home in bed, as she'd run the risk of giving birth to an idiot child. If a tree bears no fruit, it is because it was asleep during the eclipse and the Moon ate them all. So one must awaken the tree by striking it and shouting, "Help it! Help it! Bear flowers!

*

PAGES 146–152
Angelo Lomeo. *Moon Eclipse: September 26, 1996.*
Black-and-white transparencies.

Bear fruit!" One must also arouse men from their sleep, so when Râhu arrives to catch the Sun or the Moon, the monks recite prayers, music is played, cannons are fired, lanterns are lit. At the Royal Palace in Phnom Penh during the reign of King Norodom (1860-1904), the monks ordered drums, bells, and gongs to be struck in all of the pagodas, and the villagers beat on everything that could make a noise, while crying "Help us!" in order to help the Moon escape from the voracious jaws of Râhu.

The spot in the sky where an eclipse begins is also a basis for prediction. For example, if it is first visible exactly in the southeast, "there is reason to fear that houses will burn down"; if in the south, "it is a sign of mounting dissatisfaction or turmoil"; if in the northwest, "many animals will die." On the other hand, "if it begins in the true north, all animals will fare well."

IN SIBERIA, an eclipse is a monster who swallows up the planets. FROM IRAN TO MOROCCO, "the gluttonous monster" is a sea monster or a dragon. IN TURKEY AND EGYPT, the monster is slightly less cruel: it merely wishes to strangle the planet.

GRIEF AND LAMENTATION IN BABYLON

TO THE BABYLONIANS, a lunar eclipse signified the death of the Moon and was the occasion for deep mourning. The

*

OVERLEAF
Glenn Vanstrum (b. 1952).
Lunar Eclipse under Delicate Arch, Arches National Park, Utah, 1993.
Double-exposure color transparency.

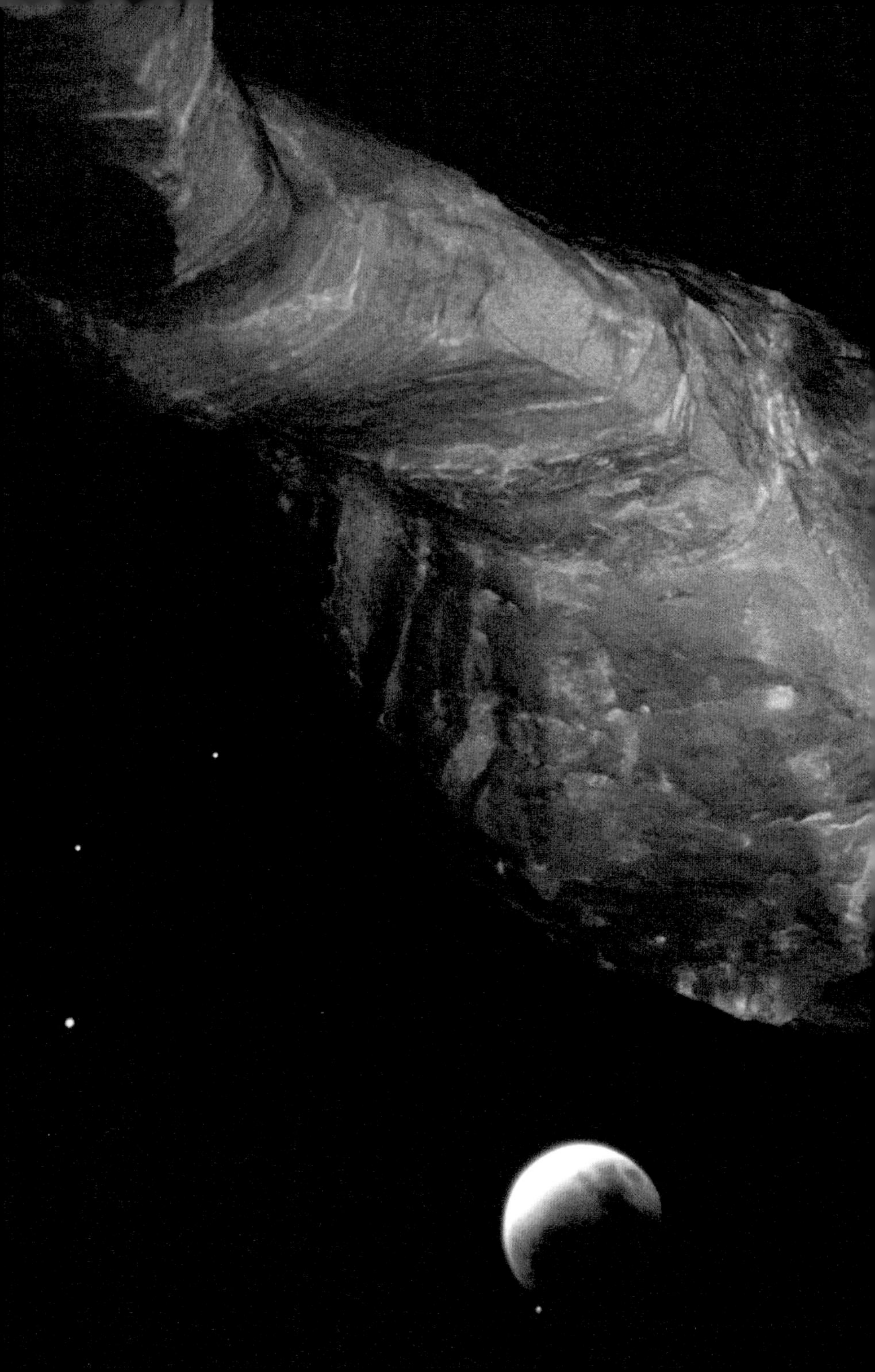

Joan Miró (1893-1983). *Dog Barking at the Moon,* 1926. Oil on canvas, $28^{7}/_{8}$ x $36^{1}/_{2}$ in. (73 x 93 cm).
PHILADELPHIA MUSEUM OF ART; A. E. GALLATIN COLLECTION.

population gathered in the town square in front of the temple gate and expressed its anguish by wailing, intoning funeral hymns, presenting offerings, tearing garments, and even making human sacrifices. Excerpts from the legend of the Evil Demon were read to help expel the captors of the Moon, accompanied by magic spells designed to free the captive. The latter were composed mostly of fire and noise, fire representing the light the Moon had lost, and noise being a sign of life, offering solace.

Noisemaking as a protective rite throughout the world

The Jews of Medina used to beat on copper urns, shouting that the Moon was bewitched. Among the Eskimos, Californian Indians, Moors, Himalayan peoples, Indonesians, Chinese, Arabs, and Persians, the same method of exorcism was employed: people clapped their hands, shouted, banged on metallic objects, shot arrows into the sky (later replaced by gunshots)–all in the hope of returning the Moon to its normal pattern of behavior.

Magic

The Moon is an accomplice in nocturnal secrets. Sometimes she hides her face, ashamed of what her light reveals. "Cynthia extinguished her fire when at its brightest, refusing to illuminate the crime" (Petronius). But most often, there is a tacit alliance between the planet of the night and nighttime lovers. "Oh Moon, linger above my bed, for it is the first time!" (Ovid).

Because of its role as an accomplice in nocturnal affairs, the Moon is associated with magic, which is most often practiced at night. And the Moon generally employs its magic to favor human love affairs. Love potions were the principal specialty of magicians in olden days, although Horace did not have a high opinion of them: "It causes me less worry and trouble to ward off thieves than to avoid those women who plague the world with their incantations and magic potions. I feel helpless to combat them, or to prevent

PAGE 158
Von Del Chamberlain (b. 1934).
American Indian Pictograph of Crescent Moon, Star and Hand Print Painted on the Ceiling of a Cliff Overhang in Chaco Canyon, New Mexico, 1979.
Kodachrome transparency.

them from gathering bones and noxious herbs as soon as the wandering Moon has shown its lovely face." Ovid was of the opinion that natural charms were more effective than magic brews: "They are far more likely to inspire love than are plants gathered by the expert hands of dangerous sorceresses. No, do not place your faith in herbs or philters."

Lowering the Moon

The art of lowering the Moon was considered to be the ultimate feat of magic in ancient times. "Magic spells can even make the Moon descend from the summit of the sky," wrote Virgil. Aristophanes mentions such an exploit in *The Clouds:* "What if I were to hire a sorceress from Thessaly, and if I made the Moon sink out of the night; if I then enclosed it in a round frame, like a mirror; and then if I kept it closely guarded?"

North African witches also knew how to lower the Moon in the sky. They would go onto a terrace or into a cemetery when the Moon was full and chant prayers, while offering a bowl of water to the moonlight. The Moon would descend from the sky in the form of foam on the surface of the water, which would start to boil—or else, some say, in the form of an angry chameleon, foaming at the mouth. They would gather the foam and raise the Moon to Heaven again—often while offering it the life of a loved one. The moon foam was then used as an ingredient in concocting love or hate potions. But it could also be used to moisten a dish of couscous that the witch had rolled in the hand of a recently disinterred corpse, and that the neglected wife would serve to her husband. Once he had eaten the couscous so specially prepared, the unfaithful husband would become as gentle as a lamb.

OPPOSITE AND ABOVE
Hieronymus Bosch (c. 1450-1516).
The Garden of Delights (exterior of triptych doors), c. 1506-16. Oil on panel, $86\frac{5}{8}$ x 153 in. (220 x 389 cm) overall.
MUSEO DEL PRADO, MADRID, SPAIN.

Vincent Van Gogh (1853-1890).
Starry Night, 1889. Oil on canvas,
29 x 36¼ in. (73.7 x 92.1 cm).
MUSEUM OF MODERN ART,
NEW YORK; ACQUIRED THROUGH
THE LILLIE P. BLISS BEQUEST.

Max Ernst (1891-1976).
Entire City, 1935-36. Oil on canvas,
$23^{5}/_{8}$ x $31^{3}/_{4}$ in. (60 x 81 cm).
KUNSTHAUS, ZURICH.

The Moon and magical healing

In human life there is not only love, but also, alas, malady. And according to whether plants are gathered during the full Moon, on the sixth day of the Moon, on the tenth day, etc., the resulting potion is supposed to bring success in love or cure of ailment. However, since these practices have more to do with folklore than with science, there is little documentary evidence of their effectiveness.

Nevertheless, Pliny the Elder, in one of the thirty-seven volumes of his *Historia naturalis,* gives a series of magic formulas concerning the Moon. For example: "It is believed that one can cause all kinds of warts to disappear by touching each one during the new Moon with a seed enclosed in a piece of cloth, which is then thrown over one's shoulder."

Superstitions and Customs

The awe inspired by Her Majesty the Moon has given rise to a vast collection of customs concerning the art and method of acquiring her good graces to best advantage, without offending her.

In Iran and Egypt, when you first glimpse the new Moon you must quickly close your eyes and open them only to see the face of your beloved; this will bring you luck.

*

OVERLEAF
Thomas Sennett (n.d.). *Moon over Mountains,* n.d. Color transparency.

John Glover (1767–1849). *Moonlight-dance of the Aborigines of van Diemen Land, Tasmania,* 1840. Oil on canvas, 30 3/8 x 57 in. (77 x 145 cm). MUSÉE DU LOUVRE, PARIS.

*

IN INDIA, a sterile woman who wishes to bear a child should bathe naked in the water, facing the Moon.

Fearful of being abducted by the Moon, women *in Greenland* refrain from drinking during the full Moon.

IN ARAB COUNTRIES, wedding nights are scheduled to coincide with the full Moon. *The Estonians, Finns, and Yakkoutes,*

*

Mask Said to Portray the Moon, collected 1909. Nisga'a, North America. Wood and pigments, 14 x 14 x 6 in. (35.8 x 35.8 x 15.3 cm). ROYAL BRITISH COLUMBIA MUSEUM, VICTORIA, BRITISH COLUMBIA, CANADA.

on the other hand, celebrate their weddings at the new Moon, symbol of fertility. *In Siberia,* too, people commence their travels and married life at the time of the new Moon.

ARABIAN NOMADS believe that the recipe for happiness is a dish of mashed peas on which moonlight has shone.

IN FEZ, MOROCCO, a moonstruck person is cured by exposure to sunlight.

IN ALASKA, during the January full Moon, the shaman wears a salmon mask to attract fish into the fishermen's nets.

THROUGHOUT THE PRE-ISLAMIC NEAR EAST, crescent-shaped objects were worn as good-luck charms.

IN ANJOU, FRANCE, warts will disappear if they are rubbed with dirt while looking at the Moon.

IN SUSSEX, ENGLAND, if the new Moon happens to fall on a Saturday, twenty days of

*

LEFT
Cubic Altar for Burning Perfume Decorated with Sun and Moon, with Inscription, 4th century B.C.
Yellow limestone, 5 5/8 x 3 3/8 in. (14.5 x 8.5 cm).
MUSÉE DU LOUVRE, PARIS.

OPPOSITE
The Sun God (top) and the God of Darkness (bottom) Making Offerings.
Page from the Codex Cospi, c. 1665.
BIBLIOTECA UNIVERSITARIA, BOLOGNA, ITALY.

Jean Cordichon, *The Properties of the Heavens.* From *Le livre des propriétés des choses,* 14th century. Manuscript.
BIBLIOTHÈQUE MUNICIPALE, REIMS, FRANCE.

*

rain and wind will follow. It should be added that there is a good chance of rain and wind in Sussex even when the new Moon falls on a Thursday.

IN CORNWALL, ENGLAND, if a child is born between the old Moon and the new Moon, he will never achieve puberty. "No Moon, no man," they say.

*

OPPOSITE
A Court Lady Leaving the Palace Grounds at Night, 2d half of the 17th century (Moghul). Watercolor on paper, 9¾ x 6⅝ in. (25 x 17 cm).
RARE BOOK DEPARTMENT, FREE LIBRARY OF PHILADELPHIA (LEWIS M151).

Where the wave of moonlight glosses
The dim grey sands with light,
For off by furthest Rosses
We foot it all the night,
Weaving olden dances,
Mingling hands and mingling glances
Till the moon has taken flight;
To and fro we leap
And chase the frothy bubbles,
While the world is full of troubles
And is anxious in its sleep.

WILLIAM BUTLER YEATS, "The Stolen Child"

PAGES 174–175
Winslow Homer (1836–1910). *Summer Night,* 1890.
Oil on canvas, $29\frac{7}{8}$ x $40\frac{1}{4}$ in. (76 x 102 cm). MUSÉE D'ORSAY, PARIS.

*

*

George Grosz (1893–1959).
Metropolis (Berlin), 1916–17. Oil on canvas, $39\frac{3}{8}$ x $40\frac{1}{4}$ in. (100 x 102 cm).
THYSSEN-BORNEMISZA COLLECTION, MADRID.

In Wales, Great Britain, if a family member dies during the night of the new Moon, three other deaths are likely to follow.

In Japan, it is highly inadvisable to look directly at the Moon on the third and sixth days of the lunar month. But a young girl who succeeds in threading a needle in the moonlight is destined to become a successful dressmaker.

Throughout the world

Never look directly at the Moon, because you run the risk of having your face turn black or of becoming blind.

If you fall asleep in moonlight, you may wake up with a grippe, a sore throat, or chronic muscular pains and headaches.

If you are in a good (or bad) mood, or if you have good (or bad) luck on the first day of the lunar month, the good or the bad will last throughout the entire month.

It is bad luck to look at the full Moon through a windowpane. If you do so by mistake, you must say: "Good evening, Madam Moon, bring me luck."

It is lucky to look at the Moon over your right shoulder. But looking at it over the left shoulder brings bad luck.

Always welcome the new Moon with a respectful greeting. If you turn a coin over in your pocket at the same time, it will bring you luck.

If lovers walk from moonlight into darkness together, they will never marry.

If a young woman in love gazes at the first Moon of the New Year through a silk handkerchief, the number of moons she sees represents the number of months before she will marry.

The Moon in Astrology

Furthermore, the Moon among the planets completes its path in the shortest time because it has the smallest orbit. It therefore passes in twenty-eight days through all the signs of the Zodiac. According to Ptolemy, under the Moon is the sign of Cancer, and Cancer is the house of the Moon.

BARTHOLOMAEUS ANGLICUS

During the first half of the sixteenth century, "predictions" were immensely popular, especially during the years preceding the conjunction of 1524, when all the planets were situated in the constellation of Pisces–an event which should have caused, according to astrologers, torrential rains, inundations and, in the opinion of some, a second Flood.

Pamphlets were written by famous astrologers, such as Nostradamus, Antoine Mizauld (the physician and fortune teller of Marguerite de Valois), Jean Thibault (doctor and astrologer of Francis I), and Jaspar Laet (professor at the University of Louvain, whose published predictions developed into a lucrative family business). Like the lunar almanacs concerned with health, these prophecies were often imitated. In Rouen, in 1557, there

PAGE 178
Sonja Bullaty. *Stonehenge,* 1969. Color transparency.

*

existed five hundred ordinary calendars indicating the days and months, and no less than seven hundred calendars of predictions!

The latter were concerned with politics, economics (the value of precious stones and gold), weather, and health. One of the first published predictions announced that the year 1475 would be warm, dry, and windy, with wars and treaties between the Jews, Saracens, Turks, and Christians, as well as epidemics of rheumatism, malaria, paralysis, and other assorted miseries. The more imaginative authors even announced the appearance of monstrous beasts, or of diseases causing horrible face swelling, among other calamities. Aside from the outlook for the entire year, these texts included monthly predictions, specifying the most favorable days for bleeding and such then-fashionable medical treatments, in which the phases of the Moon were a decisive factor.

The craze for astrology and prediction sometimes went so far as to affect the course of history. For example, the Marquis de Saluzzo, one of the favorites of Francis I of France, who regarded him with gratitude and affection, ended up by treacherously switching his allegiance to Emperor Charles V, after learning that the latter's horoscope was more auspicious. Of less importance, but no less astonishing, was the case of Alessandro Farnese, the papal legate who was sent as mediator

*

The Moon, from the Visconti-Sforza Tarot, c. 1475.
ACCADEMIA CARRARA, BERGAMO, ITALY.

PAGES 182-183
Glenn Vanstrum (b. 1952). *Moonset over Fiery Furnace, Arches National Park, Utah,* 1993. Color transparency.

Luis Alfonso Jimenez, Jr. (b. 1940).
Howl, 1977. Color lithograph on paper, 36 3/16 x 26 1/8 in. (91.9 x 66.3 cm).
NATIONAL MUSEUM OF AMERICAN ART, WASHINGTON, D.C.

between Charles V and Francis I. During a pleasant lunch with friends one day, he suddenly remembered that the favorable time for his mission, according to the planets, was about to end. He hastily left the table without so much as a good-bye, and set off at a gallop to fulfill his diplomatic mission before the planetary aspect had time to change for the worse.

Even so, astrology was not accepted as a reliable science by everyone. A humanistic movement inspired by the writings of Pico della Mirandola, a protégé of Lorenzo de' Medici, questioned its basic logic. As for the Church of Rome, it gave little credence to planetary influence on human life, for not only did it have the defect of casting doubt on man's free will, but also of almost rendering pointless the existence of God.

Some authors, such as Pietro Aretino and François Rabelais (who disliked the Louvain astrologers and called them fools), were already attacking astrology, sometimes wittily making fun of it. The author of a prediction dated 1533,

Mick Gubargu (b. 1926). *Sun, Moon and Morning Star,* 1990. Earth pigments on bark, 42¼ x 19⅝ in. (108.2 x 50.6 cm). NATIONAL GALLERY OF VICTORIA, MELBOURNE, AUSTRALIA; PURCHASE FROM ADMISSION FUNDS, 1990.

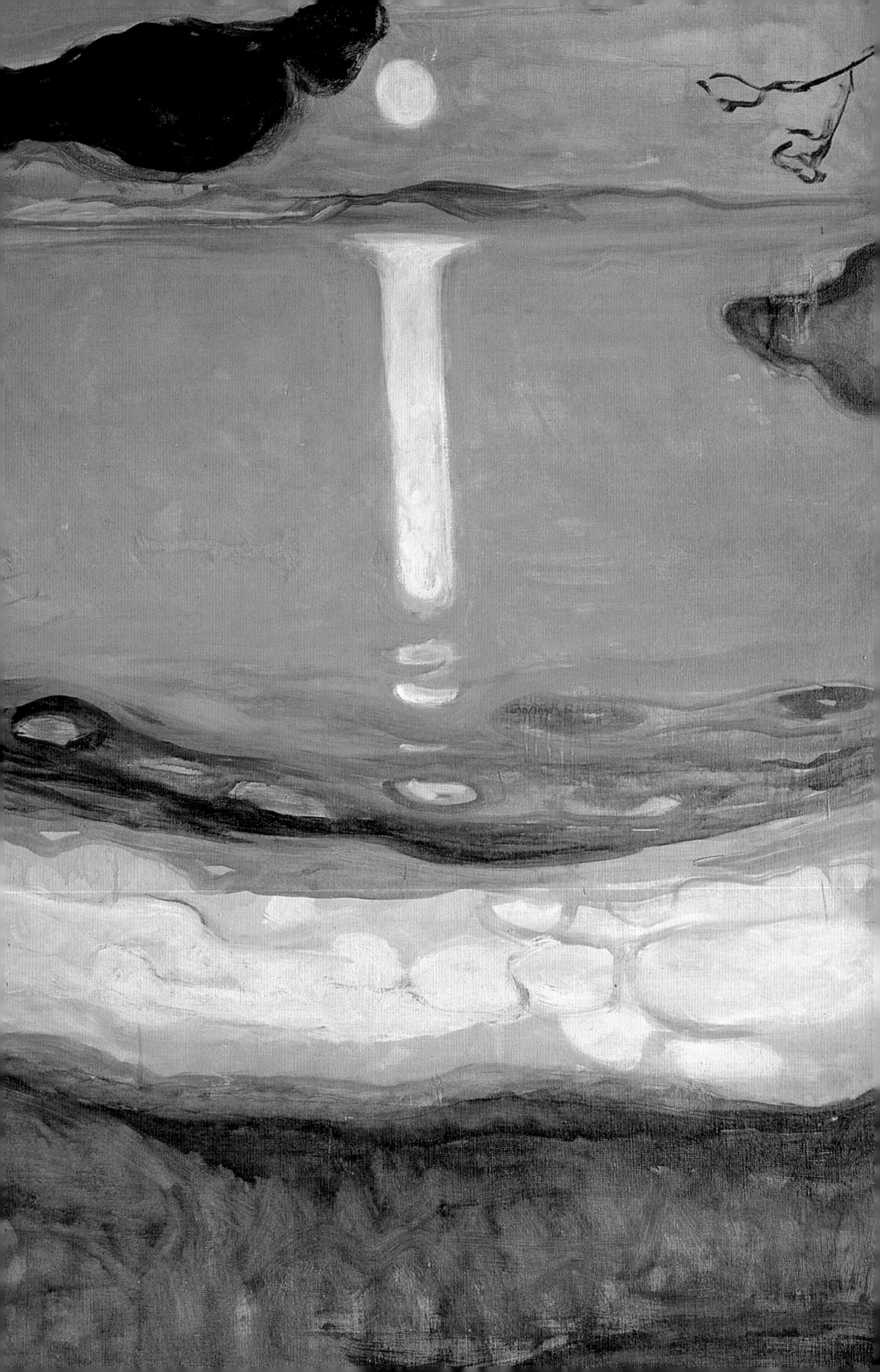

That orbèd maiden, with white fire laden,

Whom mortals call the Moon,

Glides glimmering o'er my fleece-like floor,

By the midnight breezes strewn;

And wherever the beat of her unseen feet,

Which only the angels hear,

May have broken the woof of my tent's thin roof,

The stars peep behind her and peer;

And I laugh to see them whirl and flee

Like a swarm of golden bees,

When I widen the rent in my wind-built tent,

Till the calm rivers, lakes, and seas,

Like strips of the sky fallen through me on high,

Are each paved with the moon and these.

PERCY BYSSHE SHELLEY, "The Cloud"

Edvard Munch (1863-1944).
Moonshine, 1895. Oil on canvas,
36 3/8 x 42 7/8 in. (93 x 110 cm).
NATIONAL GALLERY, OSLO.

The Meeting of Beatrice and Dante. From *The Divine Comedy,* 13th century. Manuscript. BIBLIOTECA MARCIANA, VENICE.

*

who was also opposed to the Louvain group, considering them charlatans who wrote their works "in the shadow of a glass of wine," satirized their prognostics thus: "This year pillows will be found at the foot of the bed; testicles will hang in bunches for want of game-bags; bellies will protrude; bottoms will sit down first." According to another prediction in the same vein, "This year the blind will not see much, the deaf will not hear very well, the mute will not talk a lot, the rich will be slightly better off than the poor, and healthy people will feel better than sick ones."

It was around the year 1550 that the craze for serious predictions began to decline and satirical versions to multiply. The satirists would announce that a war would end with either a victory or a peace treaty, or else "predict" in detail the events of the previous year. By the beginning of the seventeenth century, the rare predictions still published were of little influence.

However, today we can still consult astrologers, who will still draw conclusions from the sign in which the Moon was situated at the moment of one's birth. This lets us willingly submit to our impulses, no longer looking upon ourselves as reasonable, self-determined individuals, but as beings governed by instincts and inclinations.

In astrology the Moon does not symbolize the world as we observe it rationally, but is the reflection of the world as it appears to us to be–a contradiction only in conceptual terms.

The Moon symbolizes our inexhaustible reserve of psychic energy, the source of ideas and creativity; it symbolizes our acute awareness of the Infinite nourishing our dreams.

The Moon Signs

While one's sun sign in the common zodiac is determined merely by date of birth, one's moon sign is based on time of birth as well. Any astrology guide will provide tables for determining one's moon sign; below are the attributes of each moon sign.

The Moon in Aries ♈ The Fountain of Youth

THEME: Emotional extraversion

HOUSE: Zest for life, love, passion

When the Moon is in Aries, our emotions burn with unquenchable flames. Outbursts of passion, spitting flames, engulf the object of desire in glowing embers.

The Moon in Taurus ♉ The Earth Mother

THEME: Growth, sensitivity, search for emotional security

HOUSE: Plenitude and overabundance, moral warmth

Vital impulses dance like golden flames, and the soul loves so intensely that the earth boils and the sky trembles.

The Moon in Gemini ♊ A Misty Spiral

THEME: Intuitive communication, rapid comprehension of other people's feelings, seething ideas and opinions

HOUSE: Emotional wisdom, logical access to feelings and global vision due to dominant rationality

Light suddenly shines on the ever-changing reflection of the ego, disperses the structures of thought, and transforms the turbulent flow of unstable images into an ardent desire for fusion. Intellect may sometimes be undermined by emotions, and sensitivity may seem severed from psychic perceptions,

Sonja Bullaty. *Monument Valley,* 1986. Color transparency.

resulting in the divergence of soul and mind. It is this tension that engenders courage and wit.

The Moon in Cancer ♋ The Source of Creation

THEME: The mother hen; family, origins, and traditions

HOUSE: A cozy nest, retirement, the essence of past experience

In the fluid of this mysterious Moon is reflected the insatiable need of the vital waters that spring from the source of the eternal feminine. One seeks the treasure of intuition and the essence of life experience, rather than romantic adventures. The Moon in Cancer favors encounters with the primal woman (the feminine ideal). This maternal source is what Faust was seeking, and also what led Goethe to say, "The eternal feminine draws us heavenward."

The Moon in Leo ♌ Eros as a Will-o'-the-Wisp

THEME: Power and radiance, energy, and passion

HOUSE: Love and sex, creativity

The Moon in Leo represents a magnetic personality that electrifies its entourage. When love appears, there is a gigantic blaze of flames that can penetrate the very soul of animate objects, as long as the fire is constantly fed.

Elliott Erwitt (b. 1928).
Moon in the Desert, Australia, n.d.
Color transparency.

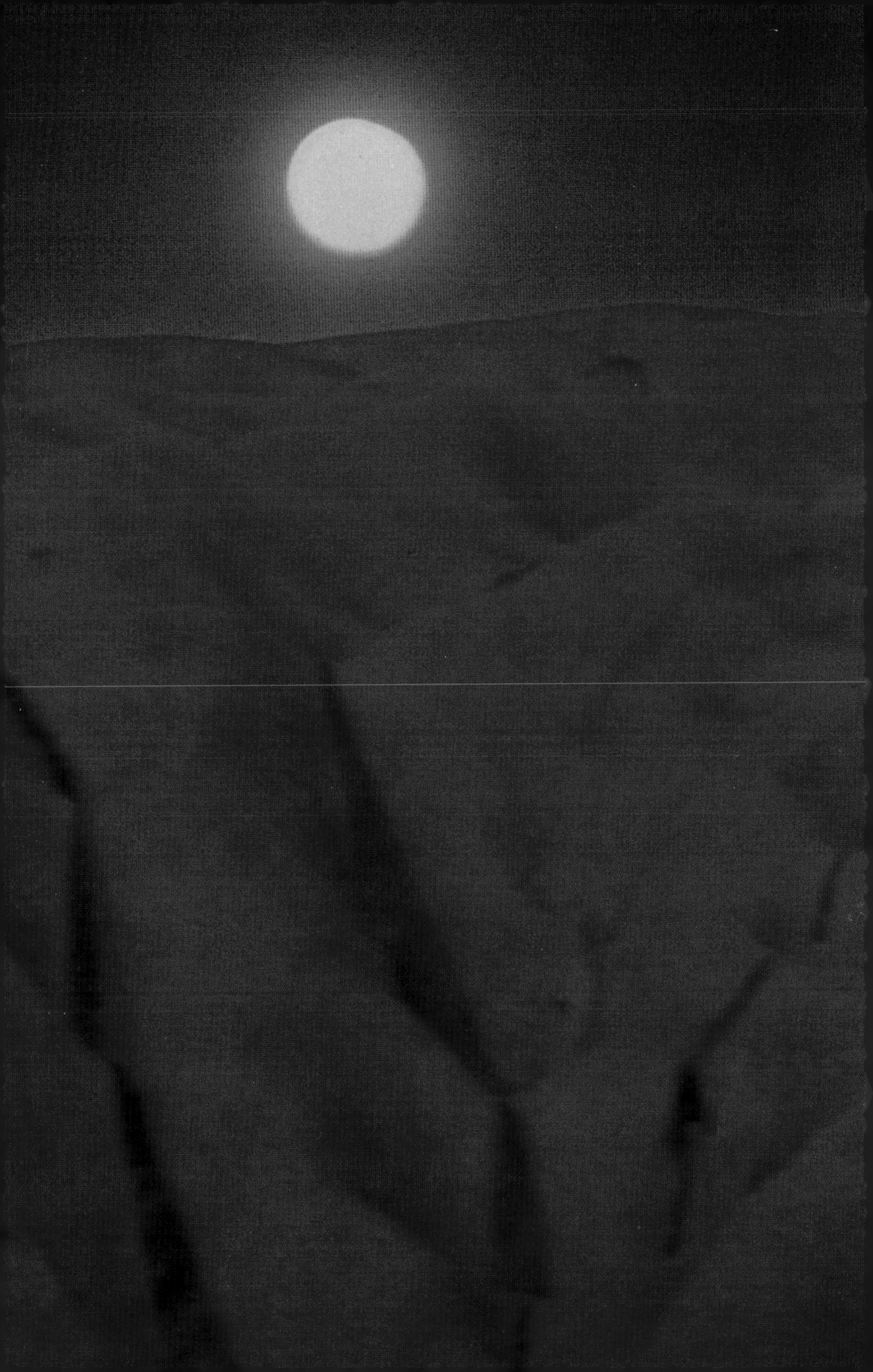

The Moon in Virgo ♍ The Salt of the Earth

THEME: Adaptability, integrity, care

HOUSE: Purification by accomplishment of duty, prudence by neutralizing emotions

Here one gladly renounces passion and remains impervious to the tumultuous effect of the Moon on the emotions. However, more than in all other signs, the body, soul, and mind seem to function in harmony, because one can view every change of circumstance with objectivity and realism.

The Moon in Libra ♎ A Bird's-Eye View

THEME: The attraction of opposites and the desire for diversity in union

HOUSE: Balance between spirit and nature

While the Sun in Libra signifies conscious conception related to reason (understanding), the Moon in Libra symbolizes moral comprehension and introspection. It expresses the attraction of opposites, nostalgia for the harmony between man and woman that was lost when they were expelled from the Garden of Eden. Since Original Sin, the union between mankind and nature has, in fact, been broken; since then the two irreconcilable camps have confronted each other. The Moon in Libra recedes into the clouds, where one has a clearer view of the pitfalls of life.

The Moon in Scorpio ♏ The Inner Sanctum of the Soul

THEME: Regeneration, transformation, contact with the dead

HOUSE: The occult, self-knowledge

The Moon in Scorpio represents the "decomposition of emotions": daughter of hell, who worships the flames of love on the altar of self-destruction. Sometimes it is lust for power, sometimes fear of Satan. The Moon in Scorpio signifies an indispensable trial by fire in order to rise from one's own ashes to explore the mysteries of life.

The Moon in Sagittarius ♐ The Grace of God

THEME: Magnanimity, goodness, social conscience

HOUSE: Intellectual and moral harmony and freedom

The Moon is influenced by Sagittarius. Driven by enthusiasm, each can set the other on fire. Images of supreme nobility and intense confusion appear as if by magic. This Moon expresses a return to the most intimate relationships.

The Moon in Capricorn ♑ A Guide to Etiquette

THEME: Caution, extreme discretion, conscientiousness

HOUSE: Self-assurance and stability

With the Moon in Capricorn, the soul hides behind armor plate in order to protect itself from attack.

I walk unseen
On the dry smooth-shaven green,
To behold the wandering moon,
Riding near her highest noon,
Like one that had been led astray
Through the heav'n's wide pathless way;
And oft, as if her head she bow'd,
Stooping through a fleecy cloud.

John Milton, *Il Penseroso*

The Moon in Aquarius ♒ THE RAINBOW

THEME: Desire to reform emotional relationships

HOUSE: Departure for unknown shores

Only gradually does one learn to interpret the reflection of psychic depths; knowledge strikes like a visionary ray of light. New perspectives loom without obscuring the whole. New perceptions shed light on past contradictions.

The Moon in Pisces ♓ THE SUBMERGED CATHEDRAL

THEME: Tact, hypersensitivity, introspection

HOUSE: Mysterious unexplored depths of the soul, quest for universal harmony

The Moon in Pisces represents a person whose principal concern is not the maternal character of nature, but instinctive inner wisdom. The Moon reflected in the waters of Pisces symbolizes a mediator possessed of high ideals and a great intuitive capacity for touching the souls of others. But in the search for unattainable ideals, there is also danger of losing one's way, since the divine path leading to the spirits of the deep in those submerged cathedrals is in reality merely a mirage.

George Robbins (b. 1933).
Rising Moon and Clouds/Powell, Wyoming, 1996.
Color transparency.

The Black Moon

The Moon moves in an elliptical orbit around the Earth, the two focal points of this ellipse being the Earth and an empty spot in space that is called the Black Moon.

The Black Moon is associated with Lilith, who, according to Cabalistic tradition, was created before Eve, at the same time as Adam, and like him directly from the Earth rather than from one of his ribs. When Lilith demanded the same status as Adam, there was a violent dispute. Lilith then fled from Eden and embarked on a demoniacal career. She thus became Eve's enemy and the instigator of illicit love.

Some astrologers consider the Black Moon insignificant, believing that one can interpret an astrological theme without taking into account an empty spot in space that isn't even a planet. Others believe that a study of the Black Moon highlights the basic problem in an individual's horoscope. Sara Sand, in *Le Grand Livre du Cancer,* states that the Black Moon symbolizes "an ordeal, a period in the wilderness, at the end of which the subject at last finds his path." For example, the Black Moon in Cancer signifies a problem concerning parents—death, disappearance, estrangement—and the subject cannot progress before having overcome this original trauma.

Be this as it may, the Black Moon represents the harmful aspects of the Moon. It symbolizes destruction, total void, sinister passions, hostile forces, vertiginous solitude, darkness to be lit. It is associated with extremes: repulsion and fascination, inaccessible spheres of thought and painful insights. If a person

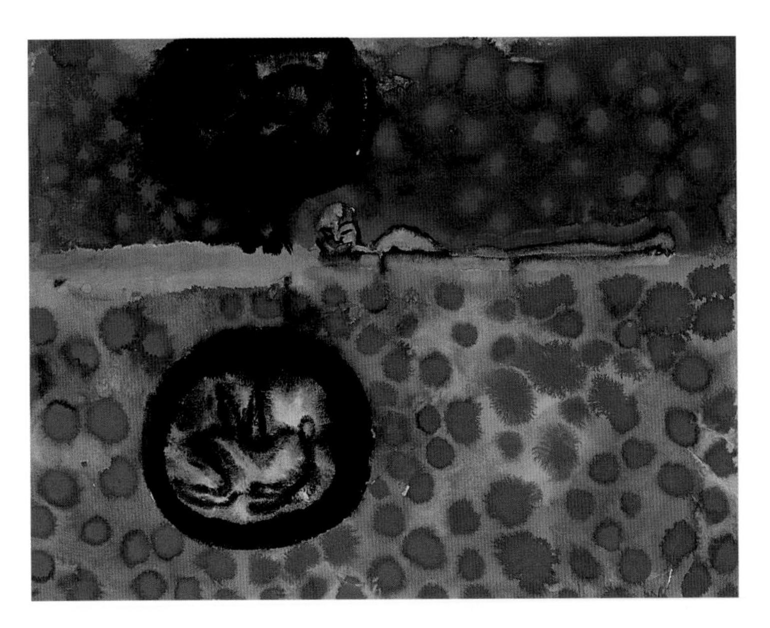

Francesco Clemente (b. 1952).
Fifty-one Days on Mount Abu: Full Moon LI, 1995.
Watercolor on handmade paper,
21 1/8 x 27 7/8 in. (54 x 71 cm).
ANTHONY D'OFFAY GALLERY, LONDON.

*

marked by the Black Moon is unable to overcome it, he prefers to withdraw from the world, even at the cost of his own destruction or that of others. But if he succeeds in transforming the poison into remedy, the Black Moon will permit him to achieve enlightenment.

What has been said about

the attributes of the Moon

and other planets suffices

for the moment.

BARTHOLOMAEUS ANGLICUS

View of Earth from Apollo 11, July 1969.
NASA photograph.

Acknowledgments

My soul thanks to Rūmī who inspired me in the first place, to my parents who showed me the beauty of the moon, and to my brother without whom I would have been half a moon.

My special thanks to Rolf Heyne, Bob Abrams, and Mark Magowan.

My day-to-day thanks to Marike Gauthier, Abigail Asher, and Ria Lottermoser.

My sincere thanks to Celia Fuller, Julien Levy, and Marie-Ange Guillaume.

My stimulating thanks to Gunter Sachs (my Beau Drac).

My astrological thanks to Kaiserli and Diana Sandmann.

My photographic thanks to Gilles Larrain.

My research thanks to Leyla Ahari, Marc Lambron, and Amir Banihashem.

My team thanks to Jutta Pakenis, Line Courtois, Stéphanie Wapler, Naomi Ben-Shahar, David Parket, Scott Hall, Hope Koturo, and Christian Diener.

And, last but not least, my best thanks to my friends and family who supported me during the project and gave me great ideas all the way through, until the moon was at its full.

—Maryam Sachs

Selected Bibliography

Cēleste. *Jardinez avec la Lune.* Paris, 1996.
Chevalier, Jean, and Alain Gheerbrandt. *Dictionnaire des symboles.* Paris, 1982.
Dupas, Alain. *La Saga de L'Espace.* Paris, 1995.
Frédērick, Robert. *L'Influence de la Lune sur les cultures.* Paris, 1978.
La Lune, mythes et rites. Paris, 1962.
Link, François. *La Lune.* Paris, 1970.
Maiello, Francesco, *Histoire du calendrier: De la liturgie à l'agenda.* Paris, 1996
Sand, Sara. *Le Grand livre du cancer.* Paris, 1998.
Vivez avec la Lune, le calendrier pour vivre au rythme de la lune. Paris 1996.
Vialatte, Alexandre. *Et c'est ainsi qu'Allah est grand.* Paris, 1979.
Watson, Lyall. *Histoire naturelle du surnaturel.* Paris, 1974.

*

Copyrights and Credits

(ARS), New York: 141; © 1998 Artists Rights Society (ARS), New York/ADAGP, Paris: 156; © 1999 Artists Rights Society (ARS), New York/ADAGP, Paris: 86; © Sonja Bullaty, New York: cover, 121, 178, 190; Rudolph Burckhardt, Courtesy of the Isamu Noguchi Foundation, Inc., Long Island City, N.Y.: 44; Chattanooga Bakery Inc., Tenn.: 40; Clore Collection, Tate Gallery, London/Art Resource, New York: 64-65; © Lee C. Coombs, Atascadero, Calif.: 77; Corbis-Bettmann, New York: 26; DC Comics, New York. Batman and all related elements are trademarks of DC Comics © 1998. Used with permission. All rights reserved: 104-5; © Von Del Chamberlain, Salt Lake City: 158; Courtesy Anthony d'Offay Gallery, London: 199; © Elliott Erwitt/Magnum Photos, New York: 36-37, 193; Werner Forman/Art Resource, New York: 168, 171; © Paul Fusco/Magnum Photos, New York: 68-69; Giraudon/Art Resource, New York: 80, 88-89, 98, 124-25, 128-29, 173, 188; © 1999 Red Grooms/Artists Rights Society (ARS), New York: 22; Herscovici/Art Resource, New York © 1998 C. Herscovici, Brussels/Artists Rights Society (ARS), New York: 8; © Hubert Josse/Abbeville Press, New York: 72-73, 97, 101, 174-75; Erich Lessing/Art Resource, New York: 20, 78-79, 94-96, 99, 136-37, 169-70; Erich Lessing/Art Resource, New York © 1999 The Munch Museum/The Munch-Ellingsen Group/Artists Rights Society (ARS), New York: 186-87; © Lin Shaozhong, Beijing, China: 32-33; © Angelo Lomeo, New York: 146-52; NASA, Washington, D.C.: 1, 201, 202; National Museum of American Art, Washington, D.C./Art Resource, New York © 1999 Luis Jimenez/Artists Rights Society (ARS), New York: 184; Nimatallah/Art Resource, New York: 176; © 1998 The Georgia O'Keeffe Foundation/Artists Rights Society (ARS), New York: 47; Photofest, New York: 43; Photofest, New York © 1999 Artists Rights Society (ARS), New York/ADAGP, Paris: 2-3; The Pierpont Morgan Library/Art Resource, New York: 28; Popperfoto/Archive Photos, New York: 48; Reuters/Corbis-Bettmann, New York: 11; © Marc Riboud/Magnum Photos, New York: 14-15, 38-39, 91, 138; © George Robbins Photo, Powell, Wyo.: 6-7, 50, 142-43, 196; Printed by permission of the Norman Rockwell Family Trust © 1967 The Norman Rockwell Family Trust, Poughkeepsie, N.Y.: 12; Courtesy of Gunter Sachs: 4, 84-85; Scala/Art Resource, New York: 16, 19, 30, 100 bottom, 132, 139, 160-61, 172, 181; © Thomas Sennett/Magnum Photos, New York: 166-67; Telimage, Paris © 1998 Man Ray Trust/Artists Rights Society (ARS), New York/ADAGP, Paris: 115; Underwood and Underwood/Corbis-Bettmann, New York: 27; © Glenn Vanstrum, La Jolla, Calif.: 154-55, 182-83; Victoria and Albert Museum, London/Art Resource, New York: 59, 123; The Andy Warhol Foundation, Inc./Art Resource, New York © 1998 Andy Warhol Foundation for the Visual Arts/Ronald Feldman Fine Arts/ARS, New York: 23.

Index of Artists and Illustrations

About the Author

Born in Iran, raised in Europe, educated in America, and residing in London with her husband and three children, Maryam Sachs is at ease in all cultures. Her first book, *The Kiss,* was published in several languages.